The publishing house tredition has created the series **TREDITION CLASSICS**. It contains classical literature works from over two thousand years. Most of these titles have been out of print and off the bookstore shelves for decades.

The book series is intended to preserve the cultural legacy and to promote the timeless works of classical literature. As a reader of a **TREDITION CLASSICS** book, the reader supports the mission to save many of the amazing works of world literature from oblivion.

The symbol of **TREDITION CLASSICS** is Johannes Gutenberg (1400 – 1468), the inventor of movable type printing.

With the series, tredition intends to make thousands of international literature classics available in printed format again – worldwide.

All books are available at book retailers worldwide in paperback and in hardcover. For more information please visit: www.tredition.com

tredition was established in 2006 by Sandra Latusseck and Soenke Schulz. Based in Hamburg, Germany, tredition offers publishing solutions to authors and publishing houses, combined with worldwide distribution of printed and digital book content. tredition is uniquely positioned to enable authors and publishing houses to create books on their own terms and without conventional manufacturing risks.

For more information please visit: www.tredition.com

Lola or, The Thought and Speech of Animals

Henny Kindermann

Imprint

This book is part of the TREDITION CLASSICS series.

Author: Henny Kindermann
Cover design: toepferschumann, Berlin (Germany)

Publisher: tredition GmbH, Hamburg (Germany)
ISBN: 978-3-8495-0852-4

www.tredition.com
www.tredition.de

Copyright:
The content of this book is sourced from the public domain.

The intention of the TREDITION CLASSICS series is to make world literature in the public domain available in printed format. Literary enthusiasts and organizations worldwide have scanned and digitally edited the original texts. tredition has subsequently formatted and redesigned the content into a modern reading layout. Therefore, we cannot guarantee the exact reproduction of the original format of a particular historic edition. Please also note that no modifications have been made to the spelling, therefore it may differ from the orthography used today.

PREFACE

It is hoped that this little work may assist in the search along the dark path upon which many a poet and—in later times—many an investigator has set his feet. It would not be worthy of us, whom science and technical ability has raised to so high an intellectual position as explorers of Nature in every field—should we neglect anything however trivial, deeming it as beneath our notice.

We know so much about all that lies around us: the manner in which the cells build our bodies; how the juices circulate within the plant. We feel Nature to be ensouled, to be a spiritual entity—and yet—it is only her corporeal life with which we are intimate. Therefore let us now turn our eyes to new horizons, so that the human spirit may be in a position to extend its search, doing so with knowledge and understanding. What is imperative is that we should investigate to what degree the higher animals have been dowered with sensibility, and to what extent this can be utilized: whether it can crystallize—so to speak—into what is known to us as *thought*. My own work of investigation was undertaken in a spirit entirely devoid of prejudice; and what I have so far discovered I now place in the hands of the reader, asking him to bring the same unbiased and objective attitude of mind to bear when reading these pages. It is my hope that they may arouse his interest and instil that broader attitude of thought which should lead to further investigation, since a question so serious and important does not permit of being lightly set aside.

I have given a short preliminary account of earlier investigations undertaken in this field of research, before inviting the reader to accompany me along the path I myself pursued into this New Land.

Henny Kindermann

CONTENTS

CHAPTER

I. THOUGHT CAPACITY IN ANIMALS
The Dogs:
Rolf
Ilse
Heinz
Harras
Roland

II. MY PREVIOUS ACQUAINTANCE WITH THE SUBJECT
III. LOLA
IV. BEGINNING THE TUITION
V. CONTINUED TUITION
VI. SENSE OF TIME
VII. CALCULATING TIME
VIII. SIGHT
IX. HER PERFECT SENSE FOR SOUND
X. SCENT
XI. SENSITIVENESS OF THE SKIN
XII. FORECASTING THE WEATHER
XIII. ADVANCED ARITHMETIC
XIV. WORKING WITH OTHER PERSONS
XV. THE QUESTION OF POSSIBLE INFLUENCE
XVI. ALTERATIONS AND MEMORY
XVII. THE CONNEXION OF IDEAS
XVIII. SPONTANEOUS REPLIES
XIX. WRONG AND UNCERTAIN ANSWERS
XX. MATTERS WHICH, SO FAR, ARE UNACCOUNTED FOR, OR UNEXPLAINED
XXI. ALTERATIONS IN CHARACTER

XXII.	A VARIETY OF ANSWERS
XXIII.	ULSE'S FIRST INSTRUCTION
XXIV.	LAST WORDS
	CONCLUSION (BY PROFESSOR H. F. ZIEGLER)
	THINKING ANIMALS (BY DR. WILLIAM MACKENZIE)

In recording the remarks made and answers given by these dogs I have—wherever it seemed possible to do so without loss of a certain distinctive charm—inserted the English translation *only*; here and there, however, where, for instance, the conversation between mistress and dog has turned on the spelling of a word it has been necessary to give the entire sentence in German. There are also some quaint remarks of which I have been loth to omit the original, these being sure to appeal to anyone acquainted with idiomatic German.

The Translator

LOLA

THOUGHT CAPACITY IN ANIMALS

It was in the year 1904 that the first experiments towards understanding an animal's ability to think were brought into public light. Wilhelm von Osten then introduced his stallion Hans II to all who seemed interested in the subject, and the most diametrically opposed opinions were soon rife with regard to the abilities of this horse, to which von Osten maintained he had succeeded in teaching both spelling and arithmetic.

The animal's mental activity was said to lie in a simple form of thinking, called into being and intensified by means of a certain amount of instruction. Von Osten, who had been a schoolmaster, had previously spent some fourteen years in testing the intelligence of two other horses before he ventured to make his experiences public, and the performances of these animals were not only remarkable, but of far-reaching importance.

Hans I, aged twelve, died in 1905. He had never appeared in public, since his abilities had been relatively modest. He had, nevertheless, been able to count up to five, as well as carry out quite a number of verbal instructions. It was Hans II, however, that convinced his master—as early as 1902—of his ability to comprehend a far greater range of the German alphabet (when written), as well as to recognize a certain number of colours.

Instances, denoting signs of evident reflection and memory, had led to Wilhelm von Osten turning his thoughts towards this work of animal tuition. Public opinion was divided; there were some who took the subject seriously and who were grateful to this innovator for thus opening a new path of inquiry; yet many were sceptical—and the scientific commission called together in 1904 to investigate the subject, finally knew no better than to heap their ridicule on the careful and patient labours of a lifetime. "Der kluge Hans" ("wise" or "clever Hans")—by that time already a public character—now evoked supercilious smiles and stood disgraced in the eyes of the majority. Only a few, capable of delving more deeply into the subject, continued to follow these performances with ever-increasing interest and amazement and kept their faith whole.

Von Osten—though now embittered and pathetically silent—quietly continued his experiments up to his death, which took place in 1909. At first he had gone about his work alone, but he was joined subsequently by Karl Krall, who then became known in connexion with this work for the first time.

Many were the attempts made in certain quarters of the Press to account for the facts of the case; the very simple means of procedure employed by von Osten were scouted and the whole thing proclaimed to be based upon trickery, influence, secret signs, an abnormal degree of training, and what not—anything and everything was seized upon in order to come into line with ordinary opinion.

Then, in the year 1905, Karl Krall, of Elberfeld, began his experiments with Hans II, encouraging, as a foundation for the furtherance of his theories, the abilities already developed in this horse, while devoting a more profound measure of insight to the entire problem.

Karl Krall, who lavished an untold amount of time and money on the question, has also raised it to an immeasurably higher plane. He has, indeed, placed a remarkable collection of carefully selected material at the service of the scientific world. With an unusual amount of devotion, backed by patience and a genuine affection for his charges, Karl Krall has carried on a work of investigation to which he assigns no narrow limits; pursuing his labours with a cheerful energy, fully convinced of the sacredness of his task.

Anyone who has come into contact with Krall must feel respect for this man, whatever doubts he may harbour as to the results obtained.

In 1908 Krall started work with two Arab stallions, Zarif and Mohammed. Both these animals learnt to count by means of rapping out the numbers with their hoofs on a board. One rap with the left fore-hoof always counted as "ten," while each rap with the right fore-hoof counted as "one" only. The number twenty-five was, therefore, composed of two left raps and five right ones. Spelling was similarly indicated by a system of raps meant to express separate letters of the alphabet. A pause followed after each number and the answers, being displayed to sight in the form of rows of numbers, it sufficed to place the letter thus indicated beneath its respective number in order to work out the reply. In the course of time these animals learnt the most varied forms of arithmetic, even to the extent of extracting the most difficult roots. They had, indeed, learnt to give answers which were, for the part, quite independent—thus supplying the most unexpected insight into their actual thinking and feeling.

They also learnt the divisions of time, while every kind of experiment was undertaken in order to test their reasoning capacity. All these attempts and the majority of results were of such a nature that it became quite impossible not to realize that further persistence along the same lines of inquiry was bound to lead to a confirmation of the assurances already given by Karl Krall with regard to his pupils' "scholarship." Many diverse opinions were heard, while the number of serious adherents to the cause as well as that of its opponents increased. Special instances to which objection had been taken on the score of supposed "influence," or of "signalling," were carefully investigated by Krall in order to clear up any implied doubts. For this purpose a blind horse, by name "Bertho," was taken in hand, proof being thus provided to confute the mythical "code of signals" supposed to exist between master and pupil. Other tests undertaken with Bertho were equally successful; Krall was, in fact, always eager and willing to submit every objection brought forward to investigation, evident though it was, that his own vast experience amply sufficed to tip the balance in his favour.

It would take us too long should we attempt to enter into any detailed discussion on this point. Krall's book, "Denkende Tiere" ("Thinking Animals")[1], may be recommended as the best source for investigation for those desiring to know more on this subject.

It must in any case be admitted that the investigations undertaken by Krall have shed a flood of light on the problem of the capacity for thought latent in our higher animals, enabling him, as we have seen, to lay down—within certain limits—in how far and in what way the existence of this capability can be *proved* where the horse is concerned. Up to the commencement of the Great War these investigations were continued, a number of different horses being used for the purpose.

In the year 1912 I became acquainted with a new contribution towards the question of animal psychology in the person of a Mannheim dog called "Rolf."

The manner in which Rolf's gifts revealed themselves was disclosed in the columns of the "Muenchner Nachrichten" as follows:

"OUR DOG ROLF

"*By Frau Paula Moekel* (née *von Moers, in Mannheim*)

"Anyone possessing an intelligent dog of his own will probably occupy himself far more with it than he is wont to do with other animals. This has been the case with our Rolf, a two-year-old Airedale terrier, which has already attained to celebrity. It was accident that led to our discovery of his talent for doing sums correctly. Our children were sitting together at work on their home-lessons, and one of my little girls—seized with a fit of inattention—was unable to solve her very easy task, viz., 122 plus 2. At length, and after the child had stumbled repeatedly over this simple answer, my patience was at an end, and I punished her. Rolf, whose attachment to the children is quite touching, looked very sad, and he gazed at Frieda with his expressive eyes as though he was anxious to help her. Seeing this I exclaimed: 'Just see what eyes Rolf is making! It looks as if *he* knew what you do not!' No sooner had I said this than Rolf, who had been lying under my writing-table, got up and came to my side. In surprise I asked him: 'Well, Rolf, do you know what two plus two amounts to?' Whereupon the animal tapped my arm with his paw *four times*—we were all speechless! After a little while we asked him again—'5 plus 5?' Here, too, the correct answer was forthcoming, and thus on the first day did we question him up to a hundred, and with equal success. After that verbal instruction became my daily occupation with the dog, in the same way that one might teach an intelligent child, Rolf entering readily into everything, indeed, we seemed to notice that his studies gave him pleasure. By degrees he became able to solve his sums correctly in every form of arithmetic, finally even getting as far as to extract two and three roots.

"We soon noticed that Rolf could also recognize letters and numerals. He read his own name easily, for when anyone began to write it on the typewriter he instantly started wagging his tail with delight. Our greatest desire now was to devise some means of communication with him and I therefore began with the following simple explanation:

"'Rolf,' I said, 'if you could say yes and no, you would be able to talk to us; now, look here! when you want to say *yes*, give us your paw *twice*, and if *no*, then give it *three* times,' and I at once put this suggestion to an easy test, for I asked him if he would like to be spanked—and he returned a decided *no*! Then I asked him if he would like some cake, to which a prompt and joyful affirmative was given. I saw therefore that Rolf understood me, and upon this mutual basis I proceeded carefully to build. At length his alphabet came into being—he having, with the exception of one or two letters, put it together entirely by himself. It was constructed thus: I would ask him, for instance, 'Rolf, how many taps with your paw are you going to give me for *a*?' and he then gave me a number which I carefully noted down. To my inexpressible pleasure I found that Rolf never forgot the numbers he had given, though I, to this day, must have my notes to hand whenever Rolf wishes to tap out anything. It is also remarkable that on a nearer investigation of his "alphabet" it becomes evident that the letters Rolf requires least are made up of the highest numbers, whereas those to which he has constant recourse have their equivalents among the lower numbers. The letters q, v, x, Rolf never uses, as though he wished to prove to me that they are entirely useless and superfluous. Rolf can recognize any money that is shown him and counts the flowers in a bunch according to their colours and varieties. He can also differentiate the high and the deep tones on any instrument, and he is even capable of telling the number of tones struck in a chord. His memory is marvellous; he remembers names and numbers over quite a period of time, once he has heard them, and he is ready to do his tasks with any persons who are sympathetic to him should he know them well enough. It is, however, difficult to get him to work as long as anyone who is not sympathetic remains in the room. What he raps out is, of course, phonetically spelt—just according to how it sounds to him, and we have not attempted to worry him with orthography! His own original remarks are delightful."

The dog, Rolf, attained in the course of time to a higher level than did the horses. This may probably be explained by the fact that dogs are, as a rule, more continuously in the company of human beings, being also due to their superior intelligence. Rolf's mode of procedure consisted in a series of raps given with his fore-paws, similar to those given by the horses with their hoofs; but Rolf used the same paw for both decimals and units, so that we had from time to time to inquire after every number rapped out—'Is it a decimal or a unit?' Whereupon he would rap 'yes,' or 'no'—as the case might be. The numbers were then written down and the answers thus obtained.

Rolf's feats of arithmetic, like those performed by the horses, included finding the square root in the most difficult problems; yet it was in the matter of spelling answers that he excelled. Indeed, he seemed to command a particularly rich vocabulary, and applied the same with the greatest accuracy and continuity, even in long answers. These replies, when collected in their proper sequence should provide us with a wealth of insight into an animal's life of feeling. Such a collection is already extant, but has not yet been made public.

Many of the dog's answers, as well as innumerable debates about him have been published in the "Mitteilungen der Gesellschaft für Tierpsychologie"[2] ("Communications of the Society for the Study of Animal Psychology"), while others may be found in the periodical "Animal Soul."[3]

Rolf has made frequent public appearances and been subjected to tests of several hours' duration. These have taken place both in the presence of his kind and gifted mistress and teacher, and also quite alone with his examiners. On every occasion of his appearance notes have been taken as to the procedures, and on one occasion these were even attested by a Notary. At such times, indeed, suggestions were not infrequently made which might be said to exceed every justifiable limit; tests were carried out prior to which the whole family had to vacate the house—carpets were taken up, in order to hunt for electric wires; window-shutters were closed; cupboards and premises searched, and sentinels posted—all this being tolerated by them with the utmost good-humour! And in spite of all this upheaval, Rolf was almost without exception ready with his replies! A fact that may well be set to his credit, when we consider how sensitive and capricious animals are by nature. Of his examiners, it may be said, that they covered themselves with confusion.

One public appearance brought him well-merited praise from a large circle of acquaintances. So excellently did he acquit himself on this occasion that I should like to place it on record.

"REPORT OF THE PUBLIC APPEARANCE OF THE SPELLING DOG ROLF

"*By Professor H. F. Ziegler*

"In order to collect subscriptions for the benefit of the Central Committee of War Charities, as also for the Society responsible for the dogs for Army Medical Service, Frau Dr. Moekel kindly consented to introduce her dog Rolf to the general public for the first time.

"The performance took place in the Hall of the Casino at Mannheim, on the 11th of May, 1914. Every seat in the Hall was taken.

"Professor Kraemer of Hohenheim opened the meeting; he dwelt on the usefulness of these dogs—trained to perform tasks in which their intelligence accounted for no small part. He alluded to the scientific importance of the new method of instruction by means of spelling—a method first brought forward in connexion with the 'Thinking Horses' belonging to Messrs. von Osten and K. Krall, and which had revealed hitherto unexpected aspects of the animal soul.

"He further pointed out the total absence of any intentional or unintentional signalling, an objection which has already been sufficiently disproved by the many singular and entirely spontaneous communications constantly made on such occasions. Finally, he emphasized that the investigations Frau Dr. Moekel had made with her dog had proved to be of immense value as contributions towards the study of animal psychology, being, in fact, of great scientific service.

"Frau Dr. Moekel was then wheeled on to the platform in her bath-chair, and Rolf seated himself by her side.

"In the first place a number of sums were set the dog which had been called out by the audience; they were as follows: $(4 \times 7 - 13) + 3 = 5$, $2 \times 10 \div 4 = 5$, $8 \times 9 \div 12 = 6$.

"When the problem 3√27 was given Rolf proclaimed the correct number '3,'—he immediately followed this, however, by spelling out: 'nid wurdsl' ('no more roots'), implying that he declined anything further to do with that form of reckoning; he had indeed, objected to 'roots' for some time past! He next proceeded to name the various persons he recognized in the assembly—the first being, 'dand, speisl basl' (Frau Dr. Speiser, aus Basel); 'glein' (a Herr Klein, whom he had not seen for two years); further, 'ogl lsr' (Herr Landsgerichtsrat Leser). When, however, he was asked by a gentleman sitting in the front row whether he knew him (the gentleman in question had sent him notes from time to time), he replied: 'lol nid wisn' (Lol doesn't know). (N.B. Rolf is in the habit of referring to himself as 'Lol.')

"In order to subject him to an unexpected test I had brought with me a box containing a 'may-bug' made of papier mâché, the inside of which was filled with biscuits. After Frau Dr. Moekel had retired from the platform I opened my box and showed it to Rolf. He pushed his nose into it, exhibiting marked interest and seemed impatient to communicate the matter to his mistress, therefore without more ado he spelt out: 'maigfr in sagdl, inn was dsm sn' (i.e. 'Maikafer in der Schachtel; innen was zu essen') (May-bug in box; inside something to eat), adding, presumably as an after-thought, 'nid gefressn' (nicht gefressen; didn't eat it!). Rolf had therefore recognised the biscuits inside the may-bug by their smell only—and was anxious that she should know that they hadn't been given him to eat! After this a gentleman in the audience asked permission to put a secret test. The object selected was shown to the dog in such a manner that his mistress had to turn aside so as not to see it. But Rolf had become obstinate and refused to name the thing, and he insisted on spelling out: 'nid, lol rgrd der wisd man': he appeared to be 'geärgert' by the 'wüste man' (worried, or vexed by the rough man)—and it may, indeed, have been that the dog sensed a certain distrust of his mistress, or that, as is often the case with other dogs, that he was reluctant to 'show off' at the request of an entire stranger. Another time, should a similar trial be contemplated, it would be wiser if the article to be named by the dog were—even if handed up by the person desirous of making the test—shown him by someone with whom he is familiar.[4]

"Gradually Rolf became tired and rapped out: 'lol bd' (i.e. Rolf bett = Rolf to bed). A pause was made during which some of Rolf's earlier communications were made public. One was his reply as to why dogs do not like cats;[5] this ran: 'lol imr hd dsorn wn sid kdsl, freigt fon wgn graln. Lol hd lib sis dsi di nid dud grdsn lol, abr, andr hundl, di nid gnn ir.' (= Lol is always angry when he sees cats, perhaps on account of their claws: Lol loves sweet Daisy, who doesn't scratch Lol—but other dogs who do not know her.)

"On 20 August, 1914, he rapped out a remark that referred to the war; it had, of course, been difficult to explain the *nature* of war to him; the only way in which it seemed at all possible to bring this to his understanding was by comparing it to the scuffling and quarrelling of dogs—on which he observed: 'lol grn (i.e. gern = likes to) raufn, mudr frbidn (i.e. Mutter verbieten = Mother forbids) abr franzos raufn mit deidsn (i.e. Deutschen), mudr soln frbidn, (i.e. Mutter soll es verbieten = Mother should forbid it), di nid dirfn (dürfen) raufe, is ganz wirsd fon di (= They should not be allowed to quarrel—it is very rough of them!).

"When the tests were resumed, Frau Dr. Moekel asked Rolf: 'What was it the man called out in the street yesterday, when you were looking out of the window?' and the dog spelt out: 'egsdrablad 5 hundrd franzos un so weidr' (= special edition 5 hundred French—and so on!). The laughter elicited by this statement appeared to offend Rolf, for he promptly spelt out the query: 'di lagn warum?' (= They laugh—why?).

"After this he applied himself to counting the flowers in a bouquet, and he was asked to whom he would like to present it. He replied: "lib adolfin" (= dear Adolphine), thus distinguishing a particular lady who was present—and he further added "gomn" (i.e. kommen = come), she had therefore to step forward and receive the bouquet in person.

"Little flags were distributed next, and Rolf was told to name the country each stood for. For the yellow and black colours he spelt out: "esdeig" (Austria), for the Turkish—'dirgig'; for the Baden flag: "baadin," while the Württemberg colours he regarded as *German*! On being shown the Bavarian flag he spelt: 'lib mudr sei fei farb!'

(i.e. die feine Farbe der lieben Mutter = the brave colours of dear mother) — Frau Dr. Moekel being of Bavarian descent.

"At the close of the meeting Rolf was told to name certain melodies, and a gentleman present whistled the beginning of the song 'O, Deutschland hoch in Ehren' — but the dog did not at once recognize the song and spelt out — 'nogmal!' (i.e. noch einmal = once more!). Then the entire song was whistled to him and he spelt: 'heldons sdurm gbraus' (i.e. Heldensturm-gebraus) and, as he liked to hear singing, he added: 'Wagd fon rein singe, bid' (= Watch on the Rhine sing, please!). The same gentleman then obliged him by whistling the 'Wacht am Rhein,' but he was not quite content, for — as he subsequently observed, 'this was not singing' (dis nid singt).

"At the close of his tests Rolf was rewarded with a cake which he promptly recognised as 'basllegrl' (Basler Leckerle = a Specialité of Bâle).

"'The Heidelberger Zeitung' commented on the performance as follows:

"'The astonishment of the audience increased with every moment, while their delight and enthusiasm at the close of this remarkable and interesting evening found vent in a storm of applause.'

"Another journal, the 'Badische General Anzeige' wrote:

"'The evening's performance must have converted many who before had been sceptical.'"

Even as there are numerous horses capable of exercising similar abilities, so too, is Rolf not a solitary example among dogs of his kind to profit by instruction. Indeed, many of his descendants are receiving tuition under the guidance of different instructors, and are giving a good account of themselves.

I will here add Professor Ziegler's Report:

"NEW REPORTS CONCERNING THE CALCULATING AND SPELLING DOG[6]

"*By Professor Ziegler*

"The descendants of the dog Rolf that have been trained by Frau Dr. Moekel,[7] are now full grown, and several of them have acquitted themselves with success. These are the bitch Ilse, the two males, Heinz and Harras, and the bitch Lola, and I here purpose to set down the latest information about these animals. It is of great importance that the various persons under whose care these dogs were trained should — though independently of each other — have made similar observations. All investigators have reported the same astonishing memory, this affording the foundation for the dogs' feats in reckoning and spelling.

"As these reports come from persons resident at different places, who neither know, nor are in communication with each other, we here have the surest proof there is no secret or trick involved in the matter."

"A. Report on the Bitch Ilse

"Concerning Ilse, of whom a clergyman is the owner, Dr. Oelhausen has already given us some details in earlier numbers of our 'Communications'.[7] He now sends me the following, which he received from Frau Dr. Moekel in the summer of last year. The reverend gentleman had left Ilse for a few hours at Frau Dr. Moekel's — as he had often done before — while he went into town to make some purchases. On this particular occasion Frau Dr. Moekel noticed that Ilse looked particularly depressed, and her little daughter, Carla, being disturbed about the dog's woe-begone air, said: 'Mummy, Ilse must be in trouble! Only see how serious she is!' So Frau Dr. Moekel asked the dog: 'Ilse, are you really sorrowful?' To which Ilse responded: 'Ja, hr hib.' (= yes, Master beating!). Frau Dr. Moekel: 'But Ilse, I am sure your master is kind to you; you are imagining!'

"Ilse: 'bd'.

"Frau Dr. Moekel: 'Bed? Ilse — have you a bed?'

"Ilse: 'Nein.'

"Frau Dr. Moekel: 'But where do you sleep?'

"Ilse: 'hols.'

"Frau Dr. Moekel: 'Ilse, you poor little dog! Have you to sleep on the wood behind the stove?'

"Ilse: 'Ja!'

"Frau Dr. Moekel: 'Then I'll tell you what to do, Ilse: you just get up on to your master's bed—he needn't have it all to himself.'

"Frau Dr. Moekel said later that she had not made this suggestion seriously, that, in fact, she had said it more to quiet Carla, and had soon forgotten all about it. But the next day the dog's master called again and complained of Ilse, saying: 'What do you think of this? Ilse is really getting unbearable—the beast got into my bed last night: there she was this morning—stretched her whole length!' And Frau Dr. Moekel had now to confess that she herself had instigated this lapse on Ilse's part.

"To this account Dr. Oelhausen has added: 'This statement has several points of interest. There is firstly the complaint about 'beatings,' and secondly the comparison drawn between her own nocturnal quarters and those of Rolf. It may also be noticed that she was very sparing of her words, using, indeed, no more than the merest 'essentials'! Then, observe the careful way in which she followed 'Mother's' advice—only getting into her master's bed after he was well asleep!'

"Another incident, the details of which were supplied to him by Ilse's master, has also been communicated to us by Dr. Oelhausen:

"'The clergyman had taken several of his village school-children for a walk, during the course of which he asked them the names of the various trees. Among these was one of which no child could tell the name. Ilse, his constant companion, was also of the party, and she now pressed forward with such marked interest that her master put the question to her too. At this Ilse started rapping and spelt out the correct name—the tree was a larch. Her master was greatly surprised at this, suggested, however, that it was probably less a matter of knowledge than of thought-transference, yet Dr. Oelhausen queries whether the dog might not have heard the name mentioned on some previous outing, and her master admits that this might have been the case.'

"We know the unfaltering tenacity with which the Mannheim dog, Rolf, remembers names, so that it would seem more reasonable to ascribe the spelling of the name to her excellent memory than to thought-transference, which would be quite as inexplicable and incomprehensible.

"To the above I may add one more incident touching Ilse, which I received from Frau Dr. Moekel on 25 May, 1915:

"'Ilse will prove valuable to us, for—though I have given her no instruction—her master has achieved the very same results with her as I have with Rolf.[8] This is what took place the other day: My dear husband went to see our reverend friend and having arrived too early for Divine Service, seated himself on a high stone in the neighbourhood of the little church and not far from the parsonage. Our friend saw my husband and came out, accompanied by Ilse, to fetch him into the house. Ilse jumped up against my husband, wagged her tail, licked him—and showed so much exuberant affection that her master was quite surprised, and asked her:

"'Do you know this gentleman?' To which Ilse replied: 'No!' adding, as though as an after-thought—'Rolf!' She had evidently scented Rolf (who is her father and of whom she is very fond) about my husband's clothes'"

"B. Report on the Dog Heinz

"A second dog, by name Heinz, who came into the possession of Mr. Justice Leser in Mannheim, has proved himself to be an excellent arithmetician, and this without ever having been worried with instruction. In the same way as Rolf he gives two raps for 'yes' and three for 'no,' while four express that he is 'tired.'

"Mr. Justice Leser reports:

"If I ask Heinz whether he will do arithmetic he invariably raps "2," even though sometimes accompanying his assent with a yawn. I am generally obliged to hold out the prospect of some reward as an inducement to do his sums. I should have preferred his rapping against some article one could hold in one's hand, or that he could be induced to "rap out" on a board setting forth the numbers, and which might be placed on the floor before him; but to neither of these alternatives will he agree, having since his earliest youth learnt to rap in the same way as Rolf does. He will, however, not only rap for me, but for any person he knows well, solving such problems as: $3 + 4 - 6$, or $\sqrt{121} + 3$, or $14/2 + 4$, or 3^2, and he seldom makes a mistake, even when the sum he may be asked merely resembles the form of arithmetic he has learnt. But he generally gives up after two or three sums and is generally distracted. He can read the figures too, and generally gives a correct solution to sums which have been written down for him and which I myself have not read. Like Rolf, he only looks at the paper sideways. He reads very reluctantly. His memory is excellent; especially quick is he at recognizing those persons again who have at any time had to do with him.'

"When I was in Mannheim on 22 March, 1916, Mr. Justice Leser was kind enough to show me the dog. I put some problems to it verbally and was able to satisfy myself as to its abilities in the matter of arithmetic. Of those then put by me I still call to mind the following: '$24 \div 3 - 3$?' Answer: '5,' and '$\sqrt{10,000} - 87$?' Answer: '13.'[9]

"C. Report on the Dog Harras

"The third dog, Harras, came into the possession of Fräulein Eva Hoffmann, of Schloss Berwartstein, near Bergzabern, and was instructed by her in spelling and arithmetic with excellent results. This lady sends the following report:

"'From the very beginning his gift for arithmetic was quite remarkable. It was enough to give him an idea of how to reckon, explaining to him the different forms of arithmetic, for the dog to learn to give the right answers to easy sums immediately.

"'Fractions, decimals, cubes and the easier forms of equation, have been set him by a stranger. With some coaching he was also able to master textual problems in this way, giving eager and glad response in the form of "yes" and "no" when it came to questioning him as to his having understood or not understood—liked or not liked the subject. He usually did his sums with evident pleasure and with amazing celerity. Spelling gave him more trouble. He could not even remember an alphabet he had himself put together, and one I invented for him he only memorized after going over it many times. He took no pleasure in putting words together and got tired very soon. Some of his original remarks are that he recognized Sunday by the "dress" I had on; also that he had dreamt of a "cow" (this after having seen one when we were out walking), and so on.

"'Remarkable is his love of truth; should he have done anything that deserves punishment, he approaches me with his head hanging down and a very dejected tail—replying to the question as to whether he deserves a whipping with a reluctant "yes," and to a further enquiry as to whether he is ashamed of himself, he responds with an emphatic "yes—yes—yes!"

"'But as is the case with children, example and precept are of far greater use than corporeal punishment, although this cannot be neglected altogether. The axiom that we evolve in accordance with the treatment meted out to us is as true in the case of an animal as it is with that of a human being, and the more this is recognized and laid to heart the shorter will be the martyrdom still inflicted upon the animal kingdom.'

"In the March of this year Fräulein Hoffmann was kind enough to communicate the following incident to me; it corroborates an earlier observation made by Frau Dr. Moekel (compare 'Communications of the Society for Animal Psychology,' 1914, p. 6, or 'The Soul of an Animal,' 1916, p. 81).

"'I was sitting in the garden reading, when I heard the sound of birds twittering over their food in a tree hard by. Harras watched them attentively for some time and I told him the names of the birds—they were jays and wood-peckers. The next morning he did not come up to my room a second time with the maid, although he can generally hardly contain himself until he has had his breakfast given him. At length, when he did appear, I asked him if he had seen the birds again, and he answered "yes"; then to my question as to their names he gave "her" and "spct" (i.e. "Häher" and "Specht" = jay and woodpecker).'"

"D. Report on the Dog Roland

"Little Roland, who received his first tuition from Frau Dr. Moekel, unfortunately came to an untimely end—owing to an accident.[10] Concerning this, Frau Dr. Moekel wrote to me in March, 1915, as follows:

"'My dear little Roland—whom we called "Guckerl" (= Peep-eyes), because of his wonderful eyes, has been run over by a motor-car. He suffered terribly for two days and died on 19 March. His death is not only a sorrow to me, but a loss to the interests of the cause we have at heart, for Roland had begun to make the most delightful remarks quite spontaneously. On the last evening before the accident, he came to me and—without having been questioned—rapped out: "Rolf ark bei (s) d arm roland" (= Rolf has badly bitten poor Roland). I was not able at the time to translate his little utterance, and it was only after his death that I remembered my notes. Then, on putting them together it transpired that Roland had been bitten by Rolf because he had chased Daisy, our kitten.

"'Roland could recognize money, stamps and bank-notes; he could count flowers and bricks, and knew all the various colours and scents as well as count tones, recognize melodies and tell the time.'

"I have not added my report made with reference to Lola to the above, the object of my book being to make the reader acquainted with this dog."

MY PREVIOUS ACQUAINTANCE WITH THE SUBJECT

I cannot remember whether it was in 1912, or earlier, or possibly even later, that I heard for the first time of Karl Krall's horses at Elberfeld. No details then reached me; only just the generalities relative to their ability to count and spell. Of their fore-runner, "der kluge Hans," I had as yet heard nothing. I had been a child when Hans had made his début, so to speak; he had then vanished and the odium which had later attached to his name was, therefore, unknown to me. I may say that I was totally unprejudiced when the news of these horses reached and, indeed, as there was but little information I did not interest myself further about the subject, although it had made a momentary impression on me. A year or two later Professor Kraemer of Hohenheim arrested public attention by his investigations respecting animals, and it was there that I heard him deliver a lecture on the horses and also the dog Rolf of Mannheim, hearing further details from him in conversation with my father[11] and myself. What I then heard interested me immensely.

Professor Kraemer was a keen advocate of this subject, but I was chary of forming any opinion without deeper investigations. The possibility of "self-expression" on the part of animals did not seem to me to be beyond the bounds of belief, even though some examples which were supposed to attest to high intelligence seemed to me a little doubtful. I tried to get more information, but was hindered at the time owing to the three years' course of studies I was then pursuing at the Hohenheim School of Agriculture, so that I was neither able to try any experiments on my own part, nor even to read Krall's great work on the subject. The entire question, therefore, remained an open one—as far as I was concerned, although my father had been to Elberfeld to see the horses, and had,—after making personal tests—come to the conclusion that everything was above-board and in accordance with what it claimed to be and that the animals really did give answers which were the outcome of their own independent thinking. In addition to this I read the public communications made by Professor Ziegler at Stuttgart, as well as also his own personal opinions.

Both these gentlemen, Professor Ziegler, as well as Professor Kraemer, were known to me only in their capacity of serious and conscientious investigators, men upon whose judgment I might safely rely, so long as my own experience did not oblige me to take up a different standpoint. And further, I skimmed over everything that the Press brought forward of an opposing nature, so that I might know *their* point of view as well.

After I had passed my Academic Examination, and taken my Diploma, I took over, some six months later, the independent management of a big estate in the Rheinland, which consisted of three hundred acres. (I was able to do this on the strength of some practical experience I had had previously in Thüringen apart from my studies.)

After a year and a half I felt sufficiently at home at the work to be able to turn my attention to such matters of interest as lay outside that of my daily work, and I now called to mind the subject of the "Thinking Horses," deciding to attempt some experiments. The approach of such a solitary season as winter seemed to me particularly suited to this attempt and I placed myself in communication with Professor Ziegler so as to hear of a likely animal. It was to be a dog, and—for preference—a relation of Rolf. Indeed, I felt sure of excellent results, should my quest meet with success. A dog is of all animals *the* one that has for generations associated most with man; its attachment is of the most intimate and the most faithful nature, so that by inheritance, as it were, it would seem to be in a greater state of "preparedness" for fulfilling man's behests. Horses, oxen, asses, pigs, and poultry, etc., are each and all, of course, accustomed to the guidance of man's hand, but—here in Europe, at all events— they live their lives apart and are not so domesticated; they cannot, therefore, form so intimate an acquaintance with man, by means of eye and ear, as can enable them to comprehend both language and gestures. For practical purposes horses would seem to come next to dogs in the matter of intelligence—more particularly Arab horses. An Arab talks to his horse as he would to a friend, and the sparkle in the eye of this animal denotes its intelligence. In the matter of actual sensibility, the ox, the ass, and other creatures have practically nothing in common with us, showing an utterly foreign type of intelligence, and one, moreover, which has—owing to the existent century-old customs of keeping them isolated in their stalls— depressed even such intelligence as was originally theirs. Creatures of the wild seem only in exceptional cases to prove amenable to training, however great their intelligence may be they cannot adapt themselves to man's control, and can as a rule only imitate, seldom revealing to us any gleam of mental alertness.

Professor Ziegler recommended a bitch which was a descendant of Rolf's and advised me to pay a visit to Mannheim. I did so, and our interview was most satisfactory. It lasted three-quarters of an hour, by which time I had assured myself that the dog could answer, even though he did not tap my hand, but rapped out his remarks on a piece of cardboard held by Fräulein Moekel. Here is the account of my visit:

"Report of Fräulein Kindermann of her
Visit to the Family of Dr. Moekel, in Mannheim,
11 January, 1916.

"After hearing much about the 'thinking animals,' more particularly about the dog Rolf, and having also with great enthusiasm read everything I could find on the subject, I became obsessed with the desire to embark on this study, forming my opinion by tests carried out myself, thus personally being in a position to approach the subject with the requisite scientific accuracy.

"The Moekels assisted my desire with kindly and ready response, placing a descendant of Rolf at my disposal, and allowing me to acquire some insight into their 'spelling-method' by watching Rolf at work. Here is the account of my visit:

"Rolf was brought into a room where there was no one beyond the family and myself. Rolf ran eagerly from one to the other and jumped up at me. Holding up a little packet of biscuits, I said to him:

"'This is what Professor Ziegler sends you from Stuttgart with many greetings, and he hopes you are good, and that you will write him a letter.'

"I saw from his glance that he understood me, but it was only after Fräulein Moekel had most earnestly 'put it to him' that he consented to rap out a reply. At first it was not easy for me to follow, for—owing probably to his reluctance—he was not "working" distinctly, but by degrees I accustomed myself to his methods, and was able to "keep count" along with the others. What he rapped out was this:

"'Lib Deigler, dank für fein gegs,[12] die geben nit gegs arm lol[13] mehr schicken; mädel is lieb, gruss von lol" (= Dear Dr. Ziegler, thanks for nice biscuits: they give no biscuits to poor Lol—send more. The girl's a dear: greetings from Lol.)'

"After this I showed him some salmon wrapped up in paper, and said:

"'See! this is what I have brought for you; what is it?' To this he did not rap out 'salmon,' as we had all expected—good as it was to the smell, but 'erst riechen' (first let me smell it). This was a ruse on his part, and one to which I succumbed, for no sooner did I hold it nearer to his nose than he snatched it out of my hand! It was, however, promptly taken from him and he was told he would have to 'deserve it' first. In the meantime a young female dog had come into the room—she answered to the name of Lola, and I asked Rolf if Lola might come with me. His reply was a most decided 'No!' I put some further questions to him, and Frau von Moers particularly asked him: 'Is Lola clever? Is Lola to learn?' to which he made answer: 'Lola is clever, but she is not to learn because of the professors'—and he actually made a face, apparently he was thinking of his own experiences. I laughed, and said:

"'Lola shall have a good time with me; she shall run about in the woods and the meadows, and play with a lot of other animals, and not have to work too long; the professors shall be sent away when Lola is tired.' This evidently pleased him, and he became very friendly to me, and on my returning to my point and asking once more whether Lola might go with me, he rapped out his answer on my hand: it was 'Yes!'

"Then I told him about an ox, who, when he didn't want to work, pretended to be dead. Rolf now got very excited, and wanted to go on rapping—first on my hand, and then on the leather-covered sofa on which I was sitting. I became rather uneasy and got him to go and rap to Fräulein Moekel, for I could then follow the raps far better. And what he now had to say referred to the deceitful ox—it was: "Hat Recht: Lol immer sagen Bauchweh!" (= Quite right of him! Lol always says he has a pain in his stomach!)

"After this I showed him another box of biscuits, with a picture of a little nigger-boy on the lid, and asked:

"'What do you see on this?'

"To which he eagerly replied:

"'Wüst schwarz Bub!' (= A wild black boy!)

"Rolf then received his reward, and I took a grateful leave of the Moekels — accompanied by little Lola.

"This experience of coming into personal contact with Rolf's powers of self-expression made a deep and lasting impression on me. In spite of all the accounts I had read and heard this living proof was almost overpowering in its utter novelty, and in the feeling of emotion that came over me, I seemed to sense that 'Souls' Unrest' that a transition from the old conception of 'unreasoning' animals to this new cognition is bound to bring with it.

"My visit had been so short that I had not been able to put any questions as to the method of instruction pursued. I had not been able to experiment personally nor get any actual advice, for Frau Dr. Moekel had died in the autumn of 1915. Yet I was by no means displeased at my state of ignorance when I came to reflect on the matter, for it enabled me to 'blaze a trail,' as it were, according to my own way of thinking, perhaps even, enabling me to arrive accidentally at similar or, diametrically opposite results!"

LOLA

Lola is an Airedale terrier, born at Mannheim on 27 January, 1914, a daughter of Rolf, and of the equally thorough-bred Jela. Both these dogs were owned by the family of a barrister, Dr. Moekel. The Airedale terrier resembles the dog we call a "Schnauzer"; it is wire-haired and of medium growth; generally with a greyish-black coat and yellow feet. Its head is covered with silky curls beneath which two bright eyes are seen. These dogs are distinguished for their alert and attentive bearing, while their excellent constitution renders them specially suitable for being trained to useful pursuits; they are at the same time not an over-bred race. Professor Heck, writing on the subject of these dogs (see "Communications of the Society for Animal Psychology"),[1] says:

"We are indebted to Herr Gutbrod of Bradford for the fact that this dog has already become fairly well distributed among us. If I have been rightly informed regarding the Airedale's history it is a crossbreed between the otter-hound and the bull-terrier, this strain having been originally obtained by the factory hands of Airedale in the North of England, who thus sought to obtain a hardy dog—one not afraid of water, and that would prove a useful assistant when out poaching either water-fowl, hares or rabbits, occasions on which it is of importance to carry out the work with as little noise as possible.

"This breed provides a favourite 'house dog'; they have proved invaluable as Army Medical Service dogs, and are friendly with children. Jocularly they are called (in Germany) Petroleum dogs (= a play on the name Airedale, as pronounced in German, i.e. 'Erdoel'").

As already said, Lola's parents were the much spoken-of Rolf, the so-called "thinking" or "speaking" dog, and Jela, no longer owned by the Moekels. Jela seems to have been an unimportant little animal, not even very affectionate as a mother. The litter Lola was dropped at consisted of twelve pups; of these one died at once, and after the vicissitudes puppies are heirs to, those that remained and have become known to us, are Heinz, Harras, Ilse, and Lola. The first-named three all have their different owners by whom they are being taught with a certain amount of success—as indeed their reports have shown.

Previously to coming into my possession, Lola, had been removed from Mannheim at an early age, and had passed through many hands, undergoing, moreover, the most various attempts of instruction. Lack of time and also the war, had been answerable for these changes; twice, however, her own fidgetiness had resulted in her being deemed unsuitable, and it was felt that the attempt had proved a failure. Even Frau Dr. Moekel, into whose hands she had finally returned is said not to have thought much of her, having only been able to get her to learn "yes" (= 2), and "no" (= 3). I mention this, because it became clear to me later on that the success of such teaching does not depend solely on the patience, the love and the attention, nor even on the ability to, or the faculty for sensing the feelings of other creatures: not on the sympathy nor yet on the calm of individual persons, but rather on *a particular person being suited to a particular dog.*

No matter how great the ability of both the individual and the dog may be, should their temperaments not be in accord—every attempt will be fruitless. For instance, I feel very sure that I could not have taught Rolf; also that I shall never be able to get a sheep-dog (I still possess) to do more than answer "yes" and "no"; also that it would be the easiest thing for me to instruct Lola's daughter Ula—and so forth. There are, in short, "winners" and "blanks" and betwixt the two, every grade of differentiation. Yet, is this not equally true in the case of teaching children? The best of teachers need not prove equally suitable to all his pupils, while some other will turn out to be exactly the right person. And this only shows us the difficulties which so frequently obstruct the path of the best-intentioned people—where investigations are concerned; obstructions which they themselves oft-times do not notice, and to which no thought is given by prejudiced persons. For with animals we come up against a more acute degree of sensitiveness than we do in a child, which, owing to certain rudiments of common sense, is able to adapt itself more easily to either teacher or investigator.

Lola had remained with the Moekels for some time after the decease of that estimable lady; it was, however, ultimately found desirable to find other homes for some of the dogs. It was about that time that my inquiry as to the possibility of procuring a descendant of Rolf reached Professor Ziegler, and he at once seconded my application. Thus Lola was kindly placed at my disposal. At first I felt some misgivings owing to the fact that the dog was already two years old, and had also passed through numerous hands, yet I determined to go to Mannheim, and my visit took place as above narrated. Lola made a most delightful impression on me, and I put few tests to my choice, for I was in a state of some excitement after all that had taken place, and therefore took her away with me joyfully. It had seemed as if I *must* do this.

It was on 11 January, 1916. She sat in the railway carriage with me, and began to howl violently when she saw Mannheim disappearing from her gaze. I tried to console her, saying: "Don't cry! You shall be quite happy with me!" It was then that Lola looked at me for the first time attentively. She quieted down and our friendship seemed sealed. She was apparently resigned to her fate; she was also doubtless aware that she had played "second fiddle" at Mannheim, and that it would, therefore, be preferable to be somewhere "on her own." That something of the kind was passing through her mind I could see—also that she was quite aware that she now belonged to me, and imagined she would be alone with me. This latter surmise became evident as soon as we reached my home where the sheep-dog I had had for two years rushed out to welcome me.

Then Lola gazed at me with horror and disappointment; the reproach in her eyes was such that I could not but understand, and then—the two dogs flew at each other, for, in the meantime the sheep-dog had begun to understand too! This was remarkable, for male and female dogs do not as a rule fall foul of each other. For days I kept them apart in separate rooms, for the mere sight of each other occasioned deep growls—indeed, my position had become distinctly uncomfortable. Then I suddenly remembered having heard that if two dogs are allowed to come together—without their master being present, they will generally get to agree. I therefore hastily shut them both into one room, and went out into the fields!

When in the course of an hour's time I came home again, each dog was reposing in a corner—the image of peace; there was no further fracas, and there has never been any trouble since. Later on, indeed, both became good friends, and often played together, but it was a risky experiment and grim forebodings had beset me on that walk! But having occasion to apply the same cure in another case, I met with the same success again.

BEGINNING THE TUITION.

Lola had been four days with me—accompanying me through the house, and about the farm, at first on a lead, but soon without. Her extreme animation verged on wildness; I was struck with her elastic temperament and her constant attentiveness, and it seemed to me that this dog would hardly be able to sit still for five minutes. She already knew "yes," and "no," and in my joy at possessing a dog able to answer me, I put so many questions to her that I began to be afraid I might do her some injury. I was, in fact, so afraid, so in doubt as to my understanding, and so alive to my responsibilities in the matter, that I often wished I had not accepted the dog at all. I did not even know whether I could "teach"—much less whether I could "teach a dog," whom, moreover, no hereditary "urge" would induce to attend school once she knew that this would mean having to work and be attentive!

Doubts as to whether the dog understood me; in what way she understood me; what sort of creature a dog really was—whether she could "think," "feel," or even whether she was capable of hearing in the same way as we hear; able to see in the same way that we see with our eyes; whether she already possessed some cognition of the human language, and whether this possessed any meaning for her? For all at once I *knew* that I *knew nothing*. That I had not even the least idea as to the best manner to assume, whether I ought to be gentle or strict—these are but a few of the difficulties I found myself beset by. I was, in short, almost in despair. How could I presume to form an opinion, supposing that, merely to my own shortcomings, the animal remained an animal, that is—in as far as I was concerned—an "animal" in the same sense that all creatures have been, since time immemorial—according to man's opinion? How should I dare to attempt to add my contribution to man's store of knowledge in so weighty a matter without as much as knowing whether I possessed the requisite patience—a genuine gift for imparting tuition, and a sufficient measure of devotion? Above all, how could I have been so foolhardy as to have undertaken to make my investigations in connexion with a descendant of Rolf's! Indeed, my only excuse could be my intense love of knowledge, my reverence and high regard for science. Science—whose temple we may enter only when filled with intensest Will, and with pure Truthfulness vowed to the furtherance of her Service—be the results sweet or bitter, fraught with success or failure, easy or difficult, new, or along the well-worn paths. It was in *this* sense that I sought to adventure—was bound to venture, for the die was cast. It was, therefore, with all the powers I could bring to my aid that I decided to embark on my quest—no matter what the attendant results might force me to acknowledge. I would disregard no test that might prove a contribution towards the solving of this new question.

Vowed to these responsibilities I sat down opposite to my dog and began. Said I to myself: She knows that she has to rap with her paws, and that rapping *twice* or *three* times does not mean the same thing; she knows, therefore, that the difference between these numbers of raps has some meaning. I then began to count to her on my fingers—at first from one to five and then back, finally taking the numbers irregularly and then holding up as many fingers as composed the number in question. To my surprise the dog was quiet and attentive, and I therefore soon continued to count up to ten. In order to enforce this lesson more I placed a row of small lumps of sugar in front of her, counting them as I did so—for it seemed to me that these might draw her attention more to the *numbers*. And I also rewarded her from time to time with a little bit for having sat so still. Then, holding up four fingers, I ventured with the question: "How many fingers do I show? Rap out the number!" And to my joy she rapped "4!" Yet, thinking this might have been accidental, I held up five and said: "Rap out this number!" and taking hold of her paw this time in order to make her tap her answer on the palm of my hand. After this I ceased my questions, for it seemed impossible that she should have comprehended so readily, but I went on just repeating the numbers to her. On the following day I also only counted, and then began questioning again, for I could not understand why she refused to look at my hands any more, and was continually yawning. Therefore, without holding out my hands, I asked her: "How many make six?" At which she gave six raps. I could hardly believe it, so I asked her: "four?" and she replied with four raps. I asked for five, and she answered correctly. I was now confident that she did understand; but what mystified me was the celerity with which her answers were given, for allowing even that she had understood, this swiftness seemed incomprehensible, and I decided to form no opinion until I had tested her with higher numbers, and should be in a position to discount the possibility of accident.

On the third day—after the preliminary counting—I got as far as ten by means of questions, and ten seemed for some days to be the limit set—calling on me to halt, as it were. This notion led me to teach the dog addition first so as by this means to get over the simple questions as to the numbers, which were always given correctly.

All this I found quite easy to do, either using my fingers or using lumps of sugar for my purpose; I was at the same time careful to speak very distinctly and to use as few complicated phrases as possible. I would say, for instance, "Look here! two fingers and two fingers are $1-2-3-4$ fingers!" But soon she ceased to follow with her eyes, so that I became disheartened and thought I had gone ahead too rapidly, or, had not roused sufficient interest; not waiting for the psychological moment, but seeking to handle the sensitive mechanism of a sentient creature too roughly. Yet—surely this could not be so, for, after all, I was but tentatively trying, and, indeed it was open to me "to try"—even if without confidence! I then said: "How much is two and five?" doing so without illustrating the question with my fingers, and the dog rapped seven! I felt a warm thrill of delight, yet I controlled my joy and proceeded with my questions, although at that moment I said to myself: "A living creature has given you a conscious answer!"

We now continued: "1 and 3?" Answer: "4." "2 and 6?" Answer: "8." This seemed to me enough for one day, and I allowed her to scamper off with a reward for her diligence; then I sat and meditated on my experience. The fact was evident: the dog had understood me—I had seen it in her eyes. She had reflected first and had then tapped the palm of my hand with unwavering certainty. I had seen the process and had felt it. Now, it is not wise to be guided by one's feelings alone—our judgment should be unbiased, and so I decided to test these facts according to reason and in every conceivable way. Yet, no one having once experienced what I had, could ever forget the sensation, for it was like the dawning of some great truth, rising suddenly before one's eyes—clear and immense. It appeared to me as some beautiful gift of life, and I was seized with a feeling of reverence for all that may yet lie undiscovered. For this new light of which I had caught the first flash, as though reflected in some bright crystal such as I might hold in my hand—how I yearned to transmit it—to pass this gift—this joy—on to others as soon as the veil should have further lifted and the horizon have become wider. And, before passing on again to the practical and scientific side of these investigations, I should like to say that where we have to do with warm, pulsating life, feeling too has its rights, and must go hand-in-hand with reason. For it is feeling, love and patience that must first penetrate the *subject-matter*, while to reason is assigned the studying, the weighing and the proving along the path pursued by the creative, seeking spirit of man. Such is man: how humble by comparison is the animal! Yet should our love henceforth assign to it its own place—as well as its own rights—as our lowlier companion in the work of life.

Soon I ventured beyond ten. For lack of any more fingers I got a counting frame, such as small children use at school, and the red and white wire-strung balls assisted me to explain my meaning as plainly as I could. I had forgotten the exact manner in which such lessons had been given me, but I hoped for the best! Indeed, "logic" was part and parcel of every step taken during this course of instruction. Never having taught before, I was desperately anxious to give a logical—a reasonable—explanation of everything to this other being respecting those things which were quite clear to me. Those, too, who saw the dog was learning something new, also felt that she seemed to arrive at what I explained to her with great rapidity and by exercising thought; that, moreover, she understood the matter as I understood it, and all were convinced that there could be no doubt but that she *did think*.

I asked her, "14," "12," "15"? And the right answers were given. Then it occurred to me that with these high numbers the rapping must be an exertion, especially over a period of time, and I then called to mind about Krall's horses who had rapped out the decimals with their left hoof, and the units with their right. The next thing, therefore, was to make her understand the difference between "right" and "left." I took each paw in turn, saying "right paw!" and "left paw!" And it took her longer to remember that than I had expected, seeing how quick she had been up to the present. Yet, at length this too was accomplished and she gave each paw without mistake. Strange as it may seem, I found later on that abstract reckoning and spelling came easily enough, while the movements of any particular portion of the body—with the exception of those habitually practised—were always attended with greater difficulty. It would seem as if she understood rightly enough *with her head*, but had some trouble in translating what she understood into active motion; and this applies to all, excepting, of course, such movements as are the result of heredity, where no words, but some other incentive, such as "scent" may possibly come into play. It is difficult for human beings to grasp that there is life in the sub-conscious, and that it is in those sub-conscious regions that the will to act arises.

I now explained to her: "When you give your *left* paw *once*, it is to count as *ten*; when you give your *right* paw *once*, it is to count as *one only*. For, you see, if we go on counting there is too much work for one paw to do and it takes too long. Therefore if you want to say '12,' you must give the *left* paw *once*, and the *right* paw *twice*." I repeated this several times and then asked: "How do you rap fifteen?" And Lola rapped one (10) with the left paw and five times with the right. It was evident that she had understood me perfectly!

This gave me confidence, and that day we did additions up to twenty, all of which were successful. Indeed, the dog showed much interest in her work, and came to it readily. As a rule ten to fifteen minutes in the morning, and another quarter of an hour in the afternoon was lesson-time. As the results were generally successful, I was sometimes tempted to continue my questions for a little longer, and she would go on answering until at length she began to sigh—then I knew that she was tired. And after such extra exertion I would notice the next day both by the pupils of her eyes and her nervous trembling, that she had been over-worked—and the thought of it makes me feel ashamed, even to this day; for, was I not undertaking the whole study for the sake of animal creation, and to think that I might have been inflicting any cruelty was unbearable. And, indeed, as time went on, this did not occur again, for I kept a keener watch. Soon, too, her capabilities increased, and she was able to fulfil more easily the greater demands made on her when answering to questions. With regard to decimals and units, I made a discovery which is, I think, worth stating. The dog did not look at me, but seemed, on the contrary (on this occasion), much interested in gnawing the leg of a chair, and I thought she could not have understood me, or else she would surely have looked up at me. Yet, she had apparently only done this to cover her confusion—as it were! Indeed, this was evident from her expression, and she had heard everything right enough, for she then—and ever after—rapped her replies without "visualizing"—and I mentally returned thanks to Karl Krall for the practical advice he had given me, and which had been so opportune. Rolf rapped with one paw only, as has already been stated; one was, therefore, obliged at length to put the question to him: "1 or 10?" And Rolf would then say "yes" or "no," as the case might be. This is confusing for the onlookers, and, as a matter of fact, when I saw him at Mannheim I never knew for certain what number he had indicated. But with Krall's method of using alternate hoof or paw, any confusion or doubt is ruled out.

CONTINUED TUITION

Lola and I had now become to some extent accustomed to each other, and the daily progress assisted this mutual understanding. I felt that I had become calmer and more self-possessed, and this, too, reacted on the dog. I did my best to make the subjects interesting, and I soon had only to call her to lessons for her to scamper up to me quite eager to begin. I also attempted to make her understand that she would be able to help other dogs—in fact, help all dear animals, if she was industrious, thus showing people how much a dog could do—when it was able to count and spell! I told her how much kinder people would then be to animals, instead of treating them as though they were no better than wood or stone, and I instanced all Rolf could do, and told her of the good uses his abilities had been put to. And from thence forward I rewarded her for every good bit of work with either biscuits or sugar, on the principle that any creature that works is worthy of wage, since man receives either food or money. And I would here like to say that I once heard that the judges examining both Rolf and the horses had taken exception to the fact of the animals being encouraged to work by being given "rewards"; where, I wonder, is the man who will labour unrequited? There will, of course, always be exceptional individuals who will do a thing *for its own sake*—yet—after all—do not *they*, too, seek their reward? albeit in a more idealistic manner, since it will consist in the success of their undertaking.

Yet these gentlemen thought that animals ought to exhibit the ethical single-mindedness of exceptional individuals! The "mere beast"—so belittled, as a rule that it is vouchsafed less "right to the earth" than is the sole of a man's foot! How significant this may be said to be of the mental attitude in which these gentlemen sat in judgment: men, who, doubtless, considered they were doing their very utmost in the service of science!

After Lola had mastered the numerals as far as twenty I started her at simple multiplication, explaining these again on my fingers and the counting frame and here, too, I found her a ready pupil. Indeed, there really *does* seem something so very obvious in 2 and 2 things being 4 things! and we proceeded by degrees to multiply up to fifty.

I would say, for instance, over the morning coffee: "Lola, to-day the fours are to have a turn: $1 \times 4 = 4$, $2 \times 4 = 8$," and I would let her multiply with four about three times, straight on from the beginning first, and then dodging about irregularly. She usually did this without any mistake whatever, and I was now getting quite used to the celerity with which she worked. The only difficulties were in connexion with 10×3 and 10×4, where she would constantly make a slip, for then the left paw came into action, and her consciousness was not yet sufficiently concentrated on that left paw. Dogs and horses must, I imagine, have a most splendid faculty for visualizing figures — to judge from the rapidity with which they work.

It took us nine days to accomplish the multiplication table from two to ten, keeping up, of course, a repetition of what had already been learnt. This great speed is another point that often gives rise to doubts, yet it is found to be equally the case with all animals who are taught: I cannot account for it — I can merely say that it is so. I have thought at times that the reason may lie in the fact that dogs and horses have but a short span of life in comparison to man's, and therefore, a briefer period of youth wherein to acquire their stock of learning; that this might account for an animal being quicker than a child, which has ampler time and seems to need it all in order to lay a thorough foundation, since the multitude of subsequent impressions would otherwise swamp all our earliest rudimentary learning.

Lola answered splendidly. It now happened at times that I myself made mistakes and believing the fault to be hers, have said: "That is wrong!" But she was not to be put out, and stuck to her reply. Then, on going over it I would find that she was right after all!

I often put my question thus: "$7 \times 4 = ?$" and the reply would be — left paw 2, right paw 8: then: "$9 \times 3 = ?$" Answer: left paw 2, right paw 7; and again, "$6 \times 6 = ?$" Answer: left paw 3, right paw 6. How accurate a test this was might be gathered from the sure and quiet way in which she tapped the palm of my hand, first with her left paw three times, and then with the right, six. I held my hand quite flat, slantingly and immovable — there was nothing about it that could convey any sort of sign to her, otherwise she would not sometimes have rapped either less or more than I expected, as has happened both in her spelling and at her sums.

My thoughts now turned to the business of spelling and the replies to be here obtained. A total of figures from 1-40 would suffice in order to give expression to all the letters, while the same degree of comprehension of my spoken word was all I required. Then I began to tell Lola some four or five letters of her alphabet daily, questioning her as to each. Every day I repeated the lesson learnt on the previous one, and added four or five more letters. Her alphabet sounds as follows:

$\frac{a}{4}$	$\frac{e}{5}$	$\frac{i}{6}$	$\frac{o}{7}$	$\frac{u}{8}$	$\frac{au}{9}$	$\frac{ei}{10}$
$\frac{b\ \&\ p}{14}$	$\frac{d\ \&\ t}{15}$	$\frac{f\ \&\ v}{16}$	$\frac{s\ \&\ k}{17}$	$\frac{ch}{20}$	$\frac{ü}{21}$	$\frac{h}{24}$
$\frac{l\ \&\ p}{25}$	$\frac{m\ \&\ p}{26}$	$\frac{n\ \&\ p}{27}$	$\frac{r\ \&\ p}{34}$	$\frac{s\ \&\ p}{35}$	$\frac{w\ \&\ p}{36}$	$\frac{z\ \&\ p}{37}$

$$\frac{ja}{2} \qquad \frac{nein}{3}$$

It is particularly to be observed that the letters were pronounced as follows: K as k,' not as ka (= kay); H as h,' not as ha (= aitch); R as r, not as er (= ar;) L as l,' not as el: this was so as to free her "writing" of any extraneous difficulties. Rolf of Mannheim rapped out the "e" in "w" (= *vay* being the German pronunciation of "w"), as also in "g" (= *gay* being the German pronunciation of "g"); thus, if he wanted to write "wegen," he simply rapped "w g n." Now, I wanted Lola to learn to rap the entire word—"wegen," for instance, for this simplification of expression, as put into practice by Rolf, would be of no use to her in view of the method of pronunciation I was adopting with the consonants. Those who had taught Rolf understood his spelling quite as well as I in time came to understand Lola's, but with regard to their system the objection was frequently put forward (more especially by persons bent on maintaining an unfriendly attitude) that "any construction might be placed on these answers," and, I must admit, that there was some truth in this. Not that this objection could always be justified, yet there were sufficient grounds for it. The great value of Rolf's mode of expressing himself was shown in the way in which he added letter to letter in accordance with their sounds (and I doubt whether any mechanical aids or accessories would have been likely to achieve the same results), thus giving proof that he was capable of independent expression. Their system proved incidentally to have what I might call a "side value," for Lola's mode of expression, due to my own method of teaching led to quite different results—*yet on the same level*.

Lola now practised her alphabet in the morning and in the afternoon we continued multiplications; rather more slowly than at first, but we ultimately reached a hundred. New work was then added in the form of division and subtraction. She soon had this all so firmly fixed in her little head that I was able to put her to easy sums and ask: "What is 3 x 3 + 10 - 5?" The answer after a few seconds being "14." A hundred was rapped out with her left paw = ten raps.

As soon as she had mastered the entire alphabet I proceeded to contract the letters into words. I said: "Lola, now attend; you are going to learn to spell: you must rap out a word made of the letters you have learnt; now—Wald (wood or forest) is w, a, l, d," and I accentuated each letter very distinctly. "How many letters are there in this word?" I added, and the answer was "4."

"Good," I said, adding: "What is the first letter?" and she tapped in reply: "36/w"; "and the next?" "4/a"; "and then?" "25/l"; "and further?" "4/a." "Lola now listen to all the words I am going to say: essen (= to eat, also "food"), e, s, s, e, n; gut (= good), g, u, t; milch (= milk), m, i, l, c, h"; and so on. For many days I continued to name the words which lay nearest to her understanding, and each day I got her to do a little spelling, after first having divided the letters. But at the end of eight days I no longer took the words to pieces merely saying, very distinctly: "rap Ofen" (= stove), and she would tap: "7 16 5 27" = o f e n. "Rap Haus" (= house). This answer was: "24, 4, 9, 35" = h, a, u, s. Whenever she rapped I jotted down the figures in order to translate them later on into letters, for it was some time before I could sufficiently memorize their equivalents,

47

and was constantly making mistakes after Lola had become an "expert." Indeed, one's memory is easily liable to play tricks here in a way that may lead to endless confusion, for the sequence of the numbers is so at variance with what one is accustomed to.

Once I asked—by way of experiment—"What is this?" touching her nose. At first she seemed uncertain, but then came the reply: "3" = nein (no); so I said: "Lola, that is your *nose*; tap nose!" and she tapped—"27, 4, 35, 5" = nase (nose). "Good!" I said, "and what is *this*?" and I touched her eye, to which she at once replied with—"9, 17" = aug (auge = eye); she had apparently not been quite sure of what I wanted when I touched her nose.

And so we went on practising—sometimes doing too much, and this would give her a headache, but she had also learnt how to communicate this fact to me and would rap: "36, 5" = we (weh = pain, or hurt); nor was this malingering, for she worked willingly, doing so, indeed, to the utmost limits of her strength, when it would become apparent, alas! to anyone who saw her that her head was aching. This tendency to "keep going" is common to all our faithful domestic animals: more particularly is it the case with draft-animals, who will go on till they drop. There are very few that consciously resist work, or who humbug us by pretending they are ill. Yet, as I had told Rolf, we had one of these exceptions at the farm; it was an ox that would always lie down and sham dead, if not in the mood to work; he then stretched out his limbs and looked at his last gasp ... but no sooner did we leave him to himself than he was on his legs again and off to his stall. No amount of chastisement brought him to reason. And it was this immoral action that had jumped with Rolf's views when—without having been asked—he at once remarked: "Hat recht, lol sagen Bauchweh!" an excuse he is reported to have made very often of late.

I now tried to teach Lola to read the numbers, for she was thoroughly at home in all we had practised so far, so it did not seem too much of a venture. I cogitated, therefore, how best to begin; and finally I wrote on a sheet of paper as follows:

1	2	3	4	5	6
.	..	...	....	.....	......

and so on up to 10.

I then held this a few inches (40 centimetres) from her eyes and, pointing to each, said: "*One* dot looks like 1," etc. And then I wrote a 2 on a slip of paper and asked her what number it stood for. At the start this gave her a good deal of trouble, and I had to do a great deal of talking. She saw the dot right enough, but would give no attention to the figure. I helped her twice to compare the two, and then set the sheet up near the place where she usually lay, taking for granted that in the course of the day her eye would be bound to rest on it so frequently that she would probably have retained the impression by the next day. And something of this kind must have happened; for on the following morning after having gone through the explanation once more, and put the sheet aside, I wrote the figures at random all over another sheet of paper when she actually "spotted" them all—with the exception of "7," and a comparison of the two sheets soon enabled her to put this right, too. There could be no doubt but that she had really mastered her lesson, for the replies were rapped out with absolute certainty. I next attempted two-figured numerals; nor was this very difficult, for in 32, for instance, the 3 was rapped by the left—the "decimal" paw—and therefore meant "30," while the "2" was added by two raps from the right paw; in fact, she memorized this without any trouble—and for a few days we practised "reading numbers" assiduously, so as to get her perfect.

Here is an example:

20 + 14?   Answer: 34.   24 + 32?   Answer: 56.
         11 + 15 + 2?   Answer: 28

Here again the most surprising thing was the celerity with which the replies were given. I was at first inclined to *make* her look at the paper attentively, but she would merely glance over it, then came a moment of quick thought—and the answer was ready. (I propose to return to this point again in the chapter on "Seeing.")

In the course of such exercises it is no exaggeration to say that one does actually *see*, by an alteration in the eye, that the dog is thinking; the gaze is withdrawn, so to speak, as it is in the eye of a person engaged in the process of thinking; and then brightens when the result has been attained. I have often been so absorbed in contemplating this process in Lola that I have almost forgotten to continue the work we were engaged on.

As the lessons progressed it became easy to teach her to read the letters, for she now knew what it was all about, and she soon picked up the figures requisite for any given letter. Personally, I always use the Latin script for writing, and it was therefore more convenient to teach her this form rather than the Gothic, but for the sake of simplicity I made use of the small characters only. I wrote these out on a sheet of paper, taking care to make them very large, and with the equivalent figure under each — thus:

a	e	i	o	u	au	ei
4	5	6	7	8	9	10

and so on.

I then gave a short explanation and stood the sheet on the floor again — just as I had done in the case of the figures.

The next day I questioned her, taking the precaution to write out a few letters on another piece of paper, so as to be able, by comparing the two, to know what the word was at once. In a few instances the right answers were given immediately, but there was still a great deal of uncertainty. I suppose the entire alphabet at one dose had been too much for her! But I tried her again in the afternoon — going over the letters carefully, and set up the card once more, to "jog her memory." And the next morning she knew it nearly to perfection, and was able to follow with her raps such words as — h, o, l, z, (holz = wood), for I took care to separate the letters, fearing she would otherwise get confused. Whenever she seemed in doubt over some letter I had recourse to her alphabet card, and made her look it up herself.

I began to feel that the foundation for all that was most important had now been laid, and that at no distant future I should be able to ask her all kinds of questions, and my joy was great. For now the moment was at hand when I might hope to gain insight into the very being of this dog, get into touch with its thinking and its feeling — all of which was so immeasurably strange to me. Yet what I here anticipated was not to be reached in so short a span of time as had hitherto sufficed for her other studies. For the present Lola spelt out no more than I told her to, and I continued practising her diligently, for I felt sure that as long as it gave her any trouble a more lengthy answer—and more especially, a *spontaneous* one—would not be forthcoming. It had taken one month of study to accomplish all I have here set down, and I felt both grateful, happy, and not a little awed—and, indeed, I did my best to thank her by my sympathy and consideration. It was only later that I came to see my own inconsistency!

The elementary tuition, the form of which I had tentatively evolved was now at an end; and constant practice in the four modes of arithmetic, as well as in reading and spelling, kept her perfect. But it became important to make occasional experiments of longer or shorter duration; such tests might be either in support of, or in opposition to, each other, and of these I now propose to treat in the following pages, for they represent the "digest" of what had so far been learnt.

SENSE OF TIME

We often hear that dogs whose masters lead a very regular life get to know the time and the hours of the day's routine—such as walks and meals showing this by their behaviour. It might be easy to account for their intimate acquaintance with the hours of meals, since their stomach is practically their clock. But that a dog should know to a "tic" the time for his master's departure from the house—whatever the season of the year, tugging him by his coat—should he not be ready, or fetching his stick—allows of no other explanation than that of a canine sense of time.

This consideration led me to try and teach Lola our divisions of time on the clock in order to make my experiment in this direction. I took a clock on which the figures were inscribed in Arabic, and of which the dial—measuring 5 centimetres across (2 inches), was sufficiently plain to read. I then explained to her that a day and a night were divided into 24 parts: I said to her: "The day-time is light, and people can then go about, and eat and work; at night it is dark, and people and animals sleep—do you understand me?" She replied: "Yes!" (two raps). I said: "Into how many parts are the day and night divided?" and she answered: "24." "These portions," I continued, "are called hours, and one hour is again divided into sixty parts, and these are called minutes; and so as always to know what are the hours, and what are the minutes, people have made a clock—now look here: so as not to make it too big they have written only twelve hours on it and this thick little pointer goes round slowly and points to the number of the hours: now, how often must it go round in a day, if a day has 24 hours?" She replied: "2."

"You see, the little thick pointer is now pointing to *nine*, so it is 9 o'clock; what time will it be when it points to 4?" She answered: "4." "You remember that I told you that the hour is divided into 60 minutes?" "Yes." "Now—see! the big pointer goes round more quickly and points out the minutes: when *that* pointer has been round *once*, 60 minutes are gone—that means one hour. This big pointer starts at 12, and you see that there are five little strokes up to 1, and how many up to 2?" Lola rapped "10." "And where is the big pointer now?" "(At) 14." "What is 14—is it an hour?" "No." "Then what is it called?" "Minute." And after this Lola rested!

In an hour and a quarter I fetched the clock again and said: "Look! what does the little thick pointer say now?" She tapped an uncertain "no." So I explained once more and then said: "Now tell me!" and she answered this time, "50."

I stood the clock on the ground in front of her and questioned her twice more in the course of the day—correct replies being given. I also left the clock standing near her for the rest of the day, for I wanted the flight of time to become impressed on her, and her eye was bound to rest on the dial now and again during the course of the day. Her answers were invariably right now for, by way of test, I inquired: "How many minutes are there in half an hour?" And she replied: "30." And again: "How many minutes has a quarter of an hour—that is, an hour divided by 4?" And she answered: "15." She also showed much interest in all this, for she sat as still as could be, listening attentively to all my explanations. And I kept her interest alive by always telling her "what nice new things Lola would be able to learn," and at this she was visibly pleased.

The next day I made casual remarks as to the time of day out loud, and all this day's answers were equally good. I now saw that she had grasped the essentials—so that I could put the clock away, and there is not another in my rooms, the nearest being a big one standing in the kitchen which is on the ground floor. I never carry my watch, leaving it in a drawer—and generally forgetting to wind it up, so that if I do not ask, I seldom know what the time is. I have no sense of time whatever myself, so that to me it may seem either long or short—according to what I may be doing. I have always envied people who possessed this sense of absolute certainty in guessing the time—it is not a common gift. I make this remark "parenthetically" in my desire for trying to elucidate the causes which lie at the back of the "feeling for time."

On the third day after my first explanations I said to Lola in the course of the morning: "Tell me what time it is. I daresay you know without seeing the clock!" To which she answered "Yes!" "Then tell me the hour first," I said, and she rapped: "10;" "And now the minutes?" "35." I then went downstairs and found that the kitchen clock pointed to 10.30, but I was told that it was not quite exact, so I telephoned to the Post Office, and inquired the correct time—asking again in the afternoon when it was 4.17. I then said to Lola: "Tell me the hour?" "4," said she. "And the minutes?" "18." I made this test several times more, and as the replies were invariably right I could regard this experiment as successful. After this I allowed her to show off her accomplishment to various people, and as long as the novelty appealed to her Lola always told the time correctly and earned much praise. In the presence of Dr. Ziegler and others she gave a most excellent account of herself, and I frequently made practical use of her as my "timepiece." The change-over to "summer-time" created some slight confusion, but this was only temporarily, and was soon overcome. Later, however, she frequently *gave the wrong time!* — it was only the charm of novelty that spurred her on to her best endeavours!

Since then I have not questioned her as often — perhaps only once a week, and her replies have varied, some being very good. Only today (I am writing on 31 December, 1916) I asked her the time; it was very dusk, and I thought it must be nearly 5 o'clock, but Lola rapped out: "4" — "And how many minutes?" I inquired. "No!" came the reply. "Nonsense!" I cried, "there must be some minutes as well?" "No!" she insisted. So I went and assured myself, believing Lola to have been obstinate, but no, it was actually only just four!

It may be taken for granted, I presume, that all dogs have this time-sense in a greater or lesser degree, and not only all dogs, but other animals also, for there are sufficient proofs to justify this assertion. Sportsmen, in particular, will be able to furnish examples in support of the theory. That Lola was able to "tell the time" was, of course, merely a matter of tuition, this having awakened her latent consciousness, and enabled her to master the signs.

In the summer of 1916 I purchased a grey parrot with the object of further studies. This bird, being very tame, was allowed to sit on the back of my chair and enjoy a few tit-bits at meal times. I always carried him on my hand from his cage to the chair, as he would not come down from the cage — preferring to clamber about without and within. One evening I had been delayed, and did not appear as punctually as usual. My maid told me, however, that the parrot had left his cage at eight o'clock, gone straight to my chair, climbed up, and was even at that moment sitting on the back-rail waiting for me!

How sensibly animals are equipped as to the requisites of life! Probably man was, too—at one time; at a time when he stood nearer to Nature, and before his inventions and manifold accessories had weaned him from so much that was inherent and inborn knowledge.

CALCULATING TIME

At first I proposed to achieve this by building on the foundations I had already laid, on the dog's fairly reliable comprehension of the value of figures, and her knowledge of spelling. So I wrote on a large sheet of paper and in small characters:[14]

- 1 jar (jahr = year) = 365 days.
- 7 tage (= days) = 1 woche (= week).
- so for 1 jar = 52 wochen = 365 tage.

The days of the week are called: —

- 1 montag.
- 2 dinstag (dienstag).
- 3 mitwoch (mittwoch).
- 4 donerstag (donnerstag).
- 5 freitag.
- 6 samstag.
- 7 sontag (Sonntag); no work for Lola!

This was to be—at the same time—a test of Lola's reading. I placed the chart on the floor where she could look at it, and repeated: "To-morrow you must be able to know this. Now spell the first word to me. And she tapped "jar." I once more went over this new lesson, explaining it all, but put no more questions, only leaving the paper where she could from time to time look at it.

The next day I removed the chart early, and later began my questioning; fully prepared for somewhat crazy results. First I asked:

"How many days are there in a week?" She rapped "7."

"And in three weeks?" "21."

"How many weeks has a year?" "52."

I praised her warmly—her interest seemed roused, for she had rapped her answers with a sort of joyful certainty! So I continued:

"Name the second day in the week?" "dinstag!"

"And what is the day called on which you do no work?" "sontag!"

"And which day in the week is that?" "7."

I then said: "To-day is Tuesday; now remember the days carefully: to-morrow, and the day after to-morrow—and the next you must always tell me the name of the day on which I ask." I then dropped the subject, and tested her on the morrow: "What is to-day?" "Mitwoch!" I next questioned her at random as to the weeks and the year, and all her answers were correct. I was very surprised on this occasion at the short time she had taken—in spite of the rapidity of so much of her earlier work, and I began to feel a sense of certainty as to the possibility of making greater demands on her. Hitherto Lola had always been able to prove to those who have seen her at her performances that she *can* state the day of the week correctly, yet of late she has no longer taken the same delight in doing so; it has become "a bore"—and for this reason she is now only asked two or three times a month. Four days after she had learnt this accomplishment I tackled the dates. At first it was rather difficult to explain to her *why* a year, which was already divided into weeks, should be again sub-divided into months—within which, moreover, the weeks could not be disposed of in complete numbers. Once more I made out my chart, and wrote down everything as I had done on previous occasions, but with divisions into twelve parts. Then I wrote out the months and placed the number of days after each, making the addition at the bottom of the chart come to 365. I then explained to her that, besides being divided into weeks, the year was also divided into months, so that each day of the year might be more easily remembered. I told her that for instance—"this day was Saturday; that it was in the month of March, and that to-day was the 13th of March." That "yesterday had been Friday, the 12th of March, and that to-morrow would be the 14th," and so forth. Then I left my chart on the floor again, and did not refer to the subject any more that day.

On Sunday Lola was seldom given anything to do so that the divisions of the week should be firmly planted in her memory. Having, therefore, removed the chart on Sunday, I asked her on Monday:

"How many months has the year?" Answer: "12."

"And what is the second month called?" "February."

She was very eager and giving her undivided attention to the work, so I continued: "What day is to-day?" "Monday." "What number is this day?" "12." Now, this was wrong, so I said: "Yesterday was the 14th, so what is to-day?" And she replied: "15." I said: "How many days has March?" Answer: "31." This last answer seemed to me the most astonishing, especially as I had not really laid much stress on this part of the lesson—fearing I might be expecting too much from her at the beginning. As a matter of fact, I was myself by no means sure as to the number of days in March, and had to verify it first! Up to this day Lola has not forgotten how many days there are in each month, although this question has merely been asked now and again; it has not been put to her now for about nine months. Owing to the regularity of my daily work I take but little heed of dates, so it comes that I have often put the question to her, for when I *do* ask it is of importance to me to have accurate information, and I have always been able to rely on Lola's quick and steady rap, subsequent reference invariably proving that I can place implicit confidence in her.

SIGHT

A dog's sight hardly plays so important a part in canine life as do scent and hearing; yet, inferior as the eye would seem in some respects, it yet excels in others. It may be observed in the case of any dog that he only recognizes his master or any person he is acquainted with at a distance of—at most—20 metres. If either my old sheep-dog or Lola come to meet me they do not see first *at all* that there is a person standing on the road. If one moves, the dog will then recognize at a distance of some 50 metres, that a human being is in front of it—the movements being responsible for this. Then, when one gets within 10 or 20 metres, the cautious and critical aspect changes, and the dog will rush forward in joyous welcome. This is enough to show that in comparison to our sight, theirs is inferior; and there are dogs that see even much worse than in the case just cited. To test this it is well to stand against the wind, otherwise the dog scents what it cannot see. It is the same case with game. At the distance, therefore, the canine eye does not seem quick of sight, but it becomes all the sharper at close quarters. Here the swift glance and good memory far out-strip our own equipment.

It was conspicuous from the beginning—both in counting and spelling—that Lola was able to learn and memorize in a surprisingly short time. Lola's charts of figures and letters were written in my none-too-clear handwriting—and yet she could remember combinations of figures amounting to ten in number from one day to the other. She could also recognize persons from their portraits, and pictures of objects familiar to her, a faculty of observation I have tested in numerous little ways. This gift was also possessed by Krall's horses and by Rolf. People seem to have the idea that dogs do not observe much, but there is no valid reason for this. Children in their *naïveté* will show their picture-book to a dog as to a friend: "Look here!" they will cry—it is only the *exception* when it occurs to a "grown-up" to do the same.

I can only say that I have convinced myself and proved to the astonishment of many that a dog *can* recognize both the letters of the alphabet and the subject of a picture shown to it.

Not that these abilities exceed those of man, at first sight, but when the matter is probed into deeply they *do* out-strip ours in one particular, and that is in celerity. For instance, if I write three or four rows of figures, one beneath the other, doing so quickly, without making any calculation myself, and then hold the paper before Lola's eyes, so that I can look into them, I see her glance skim the figures for a second or two, she will then hang her head, in evident calculation—after which she looks out straight in front of her and raps her reply. Rarely does her glance go over the paper a second time. In early days I used to think that, before holding out my hand to receive her answer, I ought to hold her head firmly and oblige her to keep her eyes on the sheet, for it seemed to me she must needs look at it for five minutes—*at least*. But Lola always tries hard to avoid looking—so I let her have her own way, and am trying to account for the cause of this quick glance by a closer study. It was the same thing when I wrote down a question—her eye flew over the sentence in three or four seconds, and the answer was given without a second glance. People to whom I have not said anything about this have stood behind me during these tests, and have generally been more impressed by the fact of her *reading* them than by the *swiftness* with which it was done. But it is the latter that amazed me most of all, for reading she and we have in common—and is indeed so far simpler a matter that there is no reason for a dog not acquiring it—but it is the *comprehension* of what it is doing, and the *speed* with which it translates what it has seen into intelligent replies that seem to me the most surprising part of all. Another instance in connexion with what I term the "cursory glance" may throw light upon this curious ability. I had heard of the way in which Rolf was able to count the flowers in a bunch, and so—on the 16 April, 1917, I thought I would try something of the same kind with Lola. For this lesson I took a sheet of paper and peppered it with dots, without any thought at regularity.

. . . . . .

. . . .

. . . . . .

. . . .

. . . . . .

Lola's first answer after looking at it for about four seconds was "34." "Are you sure?" I asked; "tell me again." She then responded with "32." I took my pencil, scratching out each dot as I went over them—there were just 32!

As she had hesitated in the first test I thought I might have made the dots too small, so taking another bit of paper I proceeded to make dots of a larger size. "How many?" I asked again. Answer: "14." I then checked this reply and found it right. The next day I covered another sheet with dots, but this time of various sizes. Lola rapped "27." "Are you sure?" I asked. "Yes!" So I counted, and there were 23. "Count again!" I commanded. "27," said she. "Lola, I can only make them 23;" "27!" insisted this dog! I could not make out the reason for this, unless, that owing to there being some writing on the reverse side, a few marks may have shown through, and thus account for the wrong answer.

On 19 April I made an attempt with red dots, but she was tired, and rapped out first 25, then 23 and finally 19—there were 19 dots. Then I made some blue dots and she rapped "11." "Are you sure?" Again "11." And this, too, was right.

I put this test several times and it was always successful when the dots were sufficiently large and regular and did not exceed 35; also if the colour was dark—either blue or black. Later on, when I read Krall's book I found that the horses had been submitted to this test with equally good results. Professor Kraemer of Hohenheim attributes the reason for this to the fact of animals having originally lived in herds, and that their "leader" as well as the other horses always knew whether their full complement was present or not. I have had the same experience with clucking-hens. A clucking-hen with twelve chicks knows at once should one be missing, and seeks it even when it cannot utter a sound, and while all the rest of her brood are running about in such confusion that it would seem impossible to count them oneself. How animals manage to do this without a sense of figures and without words always remains a puzzle to me! Now, the measure taken by a dog's eye is almost as accurate as is its sight for near objects, and its swift glance and comprehensive eye for detail. It is true that all these tests have been put to my dog Lola *alone*, but I venture to say that these facts will be found to apply to all dogs in common, should they belong to a natural and healthy breed of animals, and not to an artificially procured variety.

As to "measuring by eye," this was a test put to her accidentally. About the beginning of June, 1917, for lack of any better idea at the moment, I determined to teach her the use of the yard measure (the metre), and without having any definite object in view. So I fetched the yard-stick and told her the names and the meaning of the divisions three times; but she seemed unable to work up any enthusiasm for the subject, and I therefore did not attempt to question her. Many duties intervened, and so I forgot the whole matter for several weeks. But on 25 July I thought it might be just as well to test her eye for measure, and this reminded me of the yard-stick. So I asked for fun: "Do you remember that I showed you the yard-stick?" "Yes!" was her prompt reply. In astonishment I continued: "How many centimetres are there to the metre?" "100!" "And how many decimetres to twenty centimetres?" "2." "And how many decimetres in two and a half centimetres?" "25." Now, for the joke of the thing, I determined to test the accuracy of her eye, for I had not yet fetched the yard-stick, and she had, in fact, not seen it for many weeks. So I pointed to the outside edge of a small picture-frame that I—at a guess—took to be about twenty-two centimetres in length. At the same time I must own that I have never exercised my judgment in this line to any very great extent. "How long is this lower edge?" I asked her, "from *here* to *here*?" (pointing): her answer was, "25." I then tested it by the stick; it was twenty-six! I pointed to a larger frame, putting the same question, she answered "50." I measured, and found it to be 75. Again I showed her a smaller picture, and she rapped "19." Then I showed her a piece of chocolate—"7" was her reply—it was seven and a half. Later on, when she was in the mood she became able to guess within *half* a centimetre at a distance of about thirty centimetres and at greater distances—up to one metre; I estimated the difference to vary from about one to ten centimetres. Of late I have not given her much practice of this kind, for from the beginning she has not cared much for it. But I have made the experiment of seeing whether she can distinguish colours in the same way we do. To make this test I daubed some of the most important colours on a sheet of paper, writing the name beneath each, and the next day I daubed the same colours on another piece of paper—but in different sequence, and without adding their names. The ready response to my questions gave further proof as Lola's good memory as well as of her perfect ability to differentiate.

I next questioned her on more practical subjects. I said: "What is the colour of the stove in this room?" at the same time looking out of the window to make sure that she knew what a "stove" was. "Green," was her answer—and quite right too, for the stove is built of green porcelain tiles. I asked her a few more questions relating to flowers and to articles in daily use until I had no further doubt as to her being competent to tell one colour from the other. Coming generations may, perhaps, laugh at these numerous tests, instead of crediting animals with this ability as a matter of course!

HER PERFECT SENSE FOR SOUND

In my quest for further tests as to canine abilities, the idea occurred to me that it might be as well to arrive at a greater degree of certainty with respect to sound, that is, inquiring into a dog's memory for sound, and their powers of differentiating one tune from another. In the case of my old dog, I had already observed many things such as inclined those to whom I had related my experiences, to be of opinion that these had to do with the dog's ear. For instance, if I had been away, and returned (either driving or on foot), conversing in low tones with another person, this dog would *scream* for joy. His voice on such occasions was of quite a special quality, and everybody about the court-yard knew that I must have already passed the tree known as the "Abend Eiche," which stands some hundred metres distant, and the dog was always at that time confined, though in the open. Our conversations on such occasions were always quiet ones, and yet the dog recognized my voice at a distance of a hundred metres. If I happened to return alone and on foot, after an absence of about two days, his cries would start when I had reached *half that distance* — therefore, at fifty metres — and Lola would then also hear my step. And here is another example — one about which I was at first doubtful, not knowing to which sense it should be attributed. I always knew from Lola when I might expect a certain friend of mine — a friend to whom, by the way, she was really more attached than to me! I used to know by the heavy raps of her tail against the floor. The room in which we would be at such times was on the second floor and lay towards the front of the house. But when those anticipatory raps began my friend was still on her way, coming by a path which lay in the rear of the house, and, moreover, she always came alone. When the dog was present she could never take me by surprise.

My next ventures were of a musical nature, as I thought it might be easiest to achieve something in this direction. Lola knew the letters that are associated with the different tones ($c, d, e, f, g, a, h^{15}, e$), having learnt these in her alphabet, so I only had to strike the keys (and I confined myself to the *white* ones, as involving fewer difficulties), telling her their names. I began by saying: "Lola, you are going to learn something quite new and very beautiful; you must listen to these sounds and tell me the names of each." Then I played the notes over several times from c to c, saying clearly and slowly: "c, d, e, f, g, a, h." Then I paused and played them over again—both the ascending and descending scale.

Then I struck "c," saying, "What note is that?" She answered "c." I struck "e," but she rapped "no." I therefore played from c to e, accentuating e in particular. "Do you know now?" I asked, and she replied, "yes: e." I struck "a," and the answer came at once, "a." This seemed enough for one day, for I wished to keep her interest fresh. So we then went over some arithmetic. The next day I played only *once* from c to c, asking the names of the notes out of their order, and Lola was right in all her replies with the exception of "h," and this she soon identified after a comparison with the other notes. I tried whether she could recognize the number of notes in a chord. First I struck two, asking her the number; she replied "2." I then struck four—and she replied "4" without any hesitation. Then I struck five together, *c* being associated with them twice. At this Lola rapped "4," so I said: "You are to tell me *every* note I strike," at the same time putting down the chord again, after which she replied "5." This had been an experiment for which I had made few preparations and I marvelled at such obvious evidences of musical comprehension. But I felt that I should nevertheless test her more closely still, and so I told my experiences to a friend, a woman composer of great professional distinction. This lady was both interested and surprised, and seating herself at the piano, she struck some notes. I placed myself so as not to see the keyboard and tried to guess their pitch, yet I have no "ear" in this way. I had in 1915 attended a course of Delcroze lessons (given at Stuttgart by Fräulein Steiner) and had tried to acquire the faculty to distinguish the basic tone of any chord given at random—for this can be acquired if one is to some extent musical, yet could I but seldom succeed. I would hover in doubt between c and d, and so on, without sensing any connexion with the other tones. Here, too, with one single note being struck I was unequal to the test, but Lola's replies were excellent, yet was it again the novelty that gave zest to the affair, for later on her answers were good only when she was inclined to take trouble. But in the beginning she had been most obviously delighted with the whole matter and leapt up at me in her joy and excitement whenever I said: "Lola, listen to sounds!" I have interested and amused many friends with this little exhibition, for it came as a surprise to many, especially as the sense of "pitch" is a comparatively rare one in most people.

SCENT

The keenness of a dog's nose is, of course, proverbial, and I have only put a few tests to Lola in this particular, yet, such as they are (proving perhaps no more than is already known) I will here set down. I put the first of these tests to her on the 17 April, 1916. I showed her a book belonging to my father and said:

"Whose book is this?" She answered—"Father!" Then I showed her a glove and she told me it was mine. On 20 April, I showed her another glove belonging to a lady who was commonly known among us as "Mama" and Lola instantly replied with—"Mama!" This was followed by an important test in the afternoon of the same day. Four ladies, who were strangers to her had come to my father's place at Hohenheim, and in helping them take off their wraps I did not particularly notice where the different articles of clothing were laid. Lola was in the room at the time, I introduced the ladies to her singly and by name and later on sent her to fetch one of the hats. She fetched it and then sat expectantly before me. "To whom does this hat belong?" I asked. The answer was: "Sibol." I then asked Fräulein Sibold who was present if it really was her hat and she said—"yes." Lola had remembered the name quite well but had left out the final "d"—an omission due to the fact that I am in the habit of "swallowing" that letter when saying the name. On 29 December, 1916, I gave Lola a biscuit and she seemed more than usually delighted with its smell—as if there was something familiar about it. "Why ever are you so pleased?" I asked, to which she replied—"Mama!" And it had actually been sent by the aforementioned lady familiarly known as "Mama." I then showed her another biscuit, saying "Is this too from Mama?" but she answered "no!" "Do you dogs always know by smell?" I said—and she rapped "yes!" On this same day another test failed owing to the impossibility of ascertaining the true name of the article in question.

I had a new jacket trimmed with fur—a variety unknown to me, it was grey and slightly woolly. Lola could simply not tear herself away from it—the smell was so fascinating. I said to her: "Tell me what is delighting you so to-day?" She replied—"*mederesf*." Unable to make any sense of the letters I set them down in writing before her and asked her if any of them were wrong; to this she replied:

"yes:" "Which?" asked I—she said: "2." (*the second*) "What should it be?" I queried; she rapped "n." "How many of these letters belong to the first word?" I continued. "2." "And to the second?" She gave a wavering six—(though it may have been *five*). So the words purported to be "ne deresf." I could make nothing of it and asked her again—"What *is* deresf?" to which she gave the explanation: "ein tir." (tier = animal) "*An animal*? but I don't know the name! have you heard of it?" "Yes!" "Have we seen this animal?" "Yes!" "Where did we see it?" "Maulburg."[16] "In the house?" "No." "In the woods?" "Yes!" "Spell the name again!" "d r e s f." "And what is n e?" "dran" (a contraction of daran = on it). "On the jacket?" "Yes!" "Then you want to say that 'dresf' is on the jacket?" "Yes...." And Lola looked at me with the most imploring eyes as though I *ought to see that she was right*—as though *I ought to know it.*

"Are you *sure* of the name?" I persisted—and she replied: "mittel."[17] Here we ended—and unfortunately I have not been able to ascertain so far what this particular variety of fur is!

There have been more recent tests of this nature, about which I do not as yet feel in a position to give a definite opinion. They may possibly come into line with the theories held by Professor Gustav Jaegar, M.D., of Stuttgart and, if so, would place the subject in a new perspective. I will now only add what has so far come to my notice accidentally:

On 4 October, 1916, I said: "Lola, do you like to smell people?" "Yes!" "All people?" "No!" "How do I smell to-day?" "Tired." "Lola," I said, "do I sometimes smell horrid?" "Arger Eifersucht!" (= great, or strong jealousy) "So you smell what I feel and when it changes?" "Yes." "With every one?" "Yes." "With horses too?" "No." "With dogs?" "Yes! yes!!"

On 5 October I asked: "Lola, do I smell the same?" "No!" "How do I smell?" "Angst" (= fear, or anxiety). She evidently meant that I was uneasy on account of the amount of work.

"Lola," I continued, "how does Betty smell?" "Nach Angst" (= of anxiety) "And anything more?" "Auch müd" (= also tired). [N.B. Betty had held out the palms of her hands to the dog.] "And anything more?" "Ja—traurig" (= yes—sad.) And I found later that this had been the true state of Betty's feelings at the time.

Lola was bright and fresh and this encouraged me to continue:

"What does Magda smell like?" "Afe." "Is that right?" "No—a f." "And what more?" "g e r e g t" "afgeregt? Isn't one letter wrong?" "Yes." "Which?" "1" "Then what should it be?" "Au." "Then you mean aufgeregt?" (excited) "Yes!"

6 October. "Lola, do I smell different to-day?" "Yes—strong" "Yes! go on?" "O w e." "We?" (weh = pain) "Like pain?" "No." "You meant like the exclamation—'O weh'?" "Yes!" "But what do I smell of?" "Of surogat" (!) The use of this word by Lola seemed to be abnormal and mysterious, so I said "I am sure you have never heard that word from me!" and she replied "No!" "Tell me the name of the surogat?" "1"—(which stands for "I will not tell!") "Tell me! for you know the word for it!" I insisted. "Yes!" "*Please tell me*?" "1"—"I will not be angry," I pleaded, "I will give you a biscuit." But Lola returned again a reluctant "1." "What is this 1 to mean, Lola—is it yes or no?" "4" (= mittel). She would not look at me and while seemingly desirous of "insinuating" something, was yet not quite ready to make a frank acknowledgment of the implication. "Lola, tell me!" I exclaimed, and she rapped "Luigen." "*Lügen*?" (lying) "Ja—nein." "Lola! I won't be angry; do I smell of lies?" "Yes." "Here at home?" "Minchen." (München = Munich.) And then it suddenly dawned on me; an hour earlier I had told the dog that I was going to Munich and that perhaps she might go with me. Yet at the same time I was by no means so sure that this could be managed, and thought therefore of taking her to Stuttgart. People may smile when they read these things—indeed I have often smiled myself, but I cannot help it if Lola chooses to give such answers! Probably the future may bring me further enlightenment! There were many more occasions on which I was able to test Lola's quick nose in taking up the scent of human beings as well as of game and also the smell attaching to different articles. I need not particularize these, for anyone possessing a dog with a keen nose may know this as well as I do—or, even better.

SENSITIVENESS OF THE SKIN

The time at my disposal has unfortunately not been sufficient to enable me to engage on any very careful tests as to the sensitiveness of Lola's skin. Yet I have made certain preliminary notes as to what I hope to do in this connexion, and have also begun with a few tentative attempts. I first tried her sensibility to various degrees of warmth by teaching her the use of the thermometer. I made a drawing of a thermometer — according to its actual size — and added principal numbers and figures and also

- at 100°, water becomes air = hot.
- at 0°, water becomes hard = cold.

and beneath this I wrote:

- from 1-100 upwards, it becomes always hotter,
- from 0-40 downwards, it becomes always colder,

and I concluded with a few more verbal elucidations, and then fetched an actual thermometer on which I made her read me the temperature of the room. The next day I repeated this lesson and she read the thermometer again. After this I tested her as to whether she could give the temperature by the "feel," as it were, or whether the impression of the temperature was associated more immediately with a sense of comfort. She has so far always given the right temperature when asked, though I should add that I have only put the question to her about twenty times — and then when she has been in good health, so that I feel that the matter has not yet been sufficiently put to the proof, and I cannot, therefore, make any very definite statements with regard to this particular faculty. But I must add, that to two questions put to her on different days, she answered that she "liked her food best at 6° of warmth!" Now this chimes with the advice given in many a book on the care of dogs; "do not give them their food too hot" — and Lola's remark reminded me of this, though I might consider that "degree of heat" practically *cool* ... yet it appeared to be what she desired. Nevertheless, this preference turned out shortly to have been erroneous and, as the result of a practical trial, Lola changed her mind and voted for anything "between 12°–16°!" Here is one more test I put with regard to her susceptibility to touch: I got someone else to trace figures with their fingers on the dog's back, placing myself so that I could not see what was being described; then I put the questions, and each time her replies tallied almost invariably. One put to her in this manner was: "2 + 3?"; and "5" was given at once. While "7 + 4?" elicited a prompt "11." Then a number was described and I said: "Twice this number makes?"; to which she replied "8," four having been traced on her back. We only tried this new test for a few days so that I can give no more exact details about it—excepting this, that on that particular day, she would only understand the figures *if inscribed in this manner on her back*! It evidently amused her immensely, and we could see that she seemed to "transfer her attention," as it were, elsewhere. But though this test had been so successful with numerals, it failed entirely with letters. This was incidentally an attempt on quite a small scale at carrying out the tests which had been successfully so put to the blind horse Bertho, by Karl Krall.

These experiments as to her susceptibility to touch, or pressure, led to one slightly different, and which cannot as yet be said to have gone beyond its initial stages. I took a set of weights of 5, 10, 20, 30, 100, 200, 400, and 500 grammes, and also others of 1 and 2 kilo, and told Lola she must learn to know how heavy a thing could be. Then I placed the weights separately between her two shoulder-blades, naming them beforehand somewhat as follows—and having first written out a chart for her which set forth in a plain and easy form what I was going to say:

125 grammes	=	1/4 lb.
250 grammes	=	1/2 lb.
500 grammes	=	1 lb.
1000 grammes	=	1 kilogramme
100 lb.	=	1 zentner

I then explained this carefully and questioned her at once:

"How many pounds are 375 grammes?" Answer: "3/4."[18] "How much are 1,000 grammes?" Answer: "2." I had intentionally refrained from putting questions as to figures that were on her chart which I had left lying before her; and after she had given her replies in accordance with the pressure she had felt between her shoulders, I tested her ability at guessing where greater differences of weight were in question. On two occasions she gave the right answers, namely "1 pound" and "2 pounds," I having put the question so as to obviate superfluous spelling. I then showed her the weights, placing them in a row before her, naming them again and saying: "Which is the heaviest?" She answered "4." As a matter of fact, the heaviest of these weights, the two-pound one, was actually standing fourth. I continued: "And now?" (I had for this question transposed the weights—unseen by Lola.) Answer: "1." Which was quite right! Then—"Where is the 100 grammes?" "3." "Where is 50 grammes?" "2," and "Where is one pound?" "5." Her answers, as will be seen, were perfect; she had learnt to understand what was expected of her in this test with great rapidity.

Indeed, more elaborate tests might have been undertaken but, unfortunately, I had little leisure at the time, and was without the assistance of any educated person who might have helped me in the work. As, however, the "spade-work" in this particular field of experiment seems now to have been accomplished, many additional and interesting details might result—given the right opportunity.

It may, perhaps, be a matter of surprise, that I should have undertaken these three separate tests, and left them in their initial stages, instead of working persistently at one in particular, and thus, maybe, putting the time to better use. The reason was the old and troublesome one which was always cropping up and causing me no little worry: *Lola's interest must not be allowed to flag*. In the course of a fortnight or three weeks, for instance, I have not dared to embark on more than *one* test, not even continuing that one for as many as five consecutive days. This is why the three tests, above narrated, followed close one upon the other, while I took care to turn Lola's attention from them in between, making her go over all sorts of sums and spelling exercises. Should I have persisted in fixing her attention I should only have defeated my true object, and made her stale for future undertakings. In fact, I only engaged in these three, by way of giving a greater sense of *completeness* to the idea, and also in order to fire the ambition of others embarking upon work of a similar nature.

FORECASTING THE WEATHER

On 2 May, 1916, at a season, therefore, when farmers are generally somewhat exercised as to the coming hay-harvest, and may well wish they had some contrivance—or knew of some method whereby they could ascertain, at all events, a few days in advance what the weather is going to be, a thought flashed into my mind. At first it raised a smile, it seemed so ridiculous and impracticable, yet there could be no harm in trying. I knew that most animals, such as birds, game, etc., sensed the approach of rain at least several hours before it began to fall. But the subject is one that has not yet come sufficiently under notice, so that we do not know whether they may not sense the atmospheric changes over an even longer period. We humans are not in a position to discover how animals come by their knowledge, we can only conclude that Nature has equipped them with more delicate "chords," so to speak, and that upon these highly strung chords she can sound a warning of her impending changes, since these, our humbler brethren, stand in more imminent need thereof. It is common knowledge that animals sense earthquakes long in advance of the actual shock, and this can only be accounted for in some such way. At the time of the earthquake in 1912, Rolf, at Mannheim, crept into a corner *several hours* before it took place, and on being questioned, replied: "Lol hat angst, weiss nid vor was." (Lol is frightened; doesn't know at what.) It was quite useless trying to get further particulars as to his fears, for an earthquake was an entirely new experience to him; at a repetition of the event his remarks would, doubtless, be of greater interest and importance. Now as the weather is a matter that concerns animals, and with which they are also familiar, I determined to see how far I could get with Lola on this subject. So I taught her as follows:

- For sun = s.
- For rain = r.
- For some rain = b (ein Bischen = a little).

and to test her in this matter, I questioned her as to the last few days—here she answered correctly. Then I began:

"What about to-day?" Lola replied: "b" (= it is raining a little). I now felt sufficiently encouraged to ask her concerning the days ahead, and received the following answers:

- For 3 May = s (sun).
- For 4 May = s (sun).
- For 5 May = b (some rain).
- For 6 May = nein (no = don't know).

I told these forecastings of Lola's to several friends who, like myself, were watching the weather with anxiety. Rightly enough! the sun shone on 3 May; on that very day therefore I continued putting my questions — and Lola again prophesied:

- For 6 May = r (rain).
- For 7 May = b (a little rain).

On the next day, 4 May, the sun shone once more — as she had said it would, and in the afternoon I asked her: "How do you come to know the weather, Lola? How do you do it?" "Raten" (guessing). In astonishment I said: "From whom have you got that word?" "Dir" (from you) "Have you heard me say it?" "Yes!" On the 5th there were a few drops of rain, and on the 6th two hours' heavy downfall, but on the 7th it was dry and sunny, so that it may be that I had taxed her powers of anticipation beyond their limit, for I had asked her far in advance of the 3rd. From time to time she then continued to give me "advance information" as to the kind of weather to expect, two days or, at most, three days were the test put, and for some time I was able to fully rely on her forecasts, and would arrange my work accordingly, being careful not to cut or mow when Lola had prophesied *rain*, etc.

One morning, the sort of day when one cannot be sure of what it means to do, rain or clear, I again sought my dog's advice! It was very important to me that the hay should be carried, while the weather was dry, but I should have preferred having it loaded up towards evening, as the carts were wanted for other work — if only I knew what to expect! Lola decided for "r" (rain) in the afternoon, so

I had the hay carried at eleven—*at three the rain began*, but my loads were saved! A long period of wet weather followed; after this had continued for a fortnight—a beautiful morning broke, fine and clear, so that every one about the farm said—"at last it's going to be fine again!" I enquired of Lola—"Will there be sun to-day?" "No!" she said: "Then tell me what the weather will be to-day?" I urged. "r." I was loth to believe her, yet, by eleven, the rain had begun again. Now all this seemed very nice, and I was quite delighted, for the importance of such accuracy in agricultural work was incalculable, but I soon found that I was "reckoning without my host!" After she had—as I have shown—gone on rapping out useful and correct replies for some time, she got sick of it, began to rap out all sorts of nonsense; indeed, I knew at once from her listless and unfriendly manner that her interest was falling off, and that the replies she was giving were false. It seemed to me, indeed, that she was doing this obstinately and on purpose, so as to put me off asking any more questions! And—if so—she certainly gained her point. The lesson of this, is that one has to bear in mind that one is not dealing with a *machine*, but with a living being—and with one that is in many respects exceedingly "unreasonable" and particularly "self-willed."

I had been devoting myself to this work for some months, and had lost some of my earlier interest, but I started again three days ago so as to have another test to set down here. Lola proved to be up to the mark again, seemed interested, and I did my best to encourage her by saying: "You *will* be pleased when you know *this*!" ... "This *is* nice!" ... "See how much more a dog knows than many a man!" and so on. And as a result she announced on 5 January, 1917.

- For 6 January = b (a little rain).
- For 7 January = r (rain).
- For 8 January = r (rain).

On 6 January, there was half a degree of cold, and snow fell later in the day. This answer was near enough, for she had not been taught "snow," yet the equivalent might doubtless be found in a little "rain," i.e. wet. On 7 January, we had a heavy fall of snow, and another on 8 January. So that this test succeeded, if we discount the snow instead of rain, a change occasioned by the colder atmosphere.

ADVANCED ARITHMETIC

As the reader will now know, Lola was already acquainted with the simpler modes of arithmetic—such as addition, subtraction, multiplication and division; and we continued practising these forms for some time, even though my mind was already busy planning other and more ambitious tests. Arithmetic had of late only been taken as a corollary to her other studies, but the time seemed to have come when further advance in this too, might be deemed desirable. Her ability to "reckon" had already proved itself of practical use in facilitating her other accomplishments, and I determined now to try and put it to a still more objective test, first of all in such simple forms as: "How many people are there here?" Answer: "7." "How many of them are women?" Answer: "6." "How many dogs are there in this room?" Answer: "1." "And who is that?" "Ich" (I). A little later I said: "Listen to me, Lola! There are thirty cows in the stalls; ten of those cows go to graze, and two cows have been killed, how many cows remain in the stalls?" Answer: "18." Then I said: "Six oxen are in the stalls—how many legs have six oxen?" Answer: "24." and so we continued, the right reply being generally given after this exercise had been repeated a few times.

In May, 1916, Lola learnt the big multiplication-table, doing so easily and quickly. She was at first slightly inaccurate in the higher numbers, for rapping out the "hundreds" with the right paw and the "tens" with the left—and then again the "ones" with the right gave her some trouble in the beginning. Yet such questions as: 3 + 14, 2 + 17, 4 + 20, were given without hesitation, since these did not come within the region of the hundreds. But in time she got used to the hundreds too—and even to thousands, and to these latter she applied her left paw, rapping the date 1916 thus: left paw 1; right paw 9; left paw 1; right paw 6.

Towards the end of May I thought I would teach her fractions, and she apparently understood what I meant, but for a beginning I could only put questions, such as: "How many *wholes* are there in 20/4, 12/4, or 11/2" etc. Indeed, I was at first at a loss as to what form of expression I should use here—so as not to come into collision with those already resorted to, thus giving rise to confusion. At first I thought it might be more convenient to let her rap out the denominator with her right paw and the numerator with her left—but I soon came to see that even with 3/16, this method could no longer be maintained. At length I let her simply rap out the numerator—then I would ask for the denominator, and let her rap this, so that in the case of 3/16 she rapped the 3 first with her right paw; then gave the denominator, i.e. 1 rap with her left paw and 6 again with her right. This mode or procedure came quite naturally to her, and so it was retained. The questions were practised in the following manner:—"How do you rap 3/8, 12/6?" etc., and I followed this up with easy exercises such as: "How much is 2/8 + 1/4?" the simplified answer being "1/2." I had, as may be imagined, already given her repeated and detailed explanations on the subject before she was capable of giving such answers as "1/2," to the above question. Simplifying was also practised separately thus: "Simplify 20/16!" Answer: "1-1/4." this being given with "1 r" (pause) "1 r" (another pause); "and the denominator?" "4 r." To anyone following her actions, the meaning would appear quite distinct. I now determined that she should add together numbers having different denominators—as, for example: 1/4 + 1/3, and here I had myself to cogitate as to how this ought to be done, for at school, my enthusiasm for arithmetic had never been great and much of what I had then learnt has been forgotten. So I talked the question over with a friend—in Lola's presence and out loud—and finally arrived at the solution. As she had been listening most of the time while we sought, found, and discussed the solution, I soon ventured to put a few tests to her, and the answers proved that she had actually been listening while our conversation was going on, and that what we had talked about had lingered in her memory. By the way, it is reported of Jean Paul Richter, that when on some occasion a friend came to him desirous of talking over some matter, the nature of which none other was to know, Jean Paul said to his poodle, who was under the table: "Go outside, we want to be alone!" The dog vacated, and the poet remarked: "Now, sir, you can talk, for no one will hear us!"

Lola solved the following problems:

"1/5 + 1/3 = ?"   A. "8/15."   "1/7 + 5/8 = ?"   A. "43/56."
"1/2 + 1/3 = ?"   A. "5/6."   "1/4 + 2/5 = ?"   A. "13/20."

As the problems always took me longer than they did her I never checked them at the time, but went over them later, after she had given all her answers. I did this moreover, so that she should have no opportunity of tapping my thoughts and thus rely on me; indeed, I really *forced* her to do her own thinking. For even if I did begin to calculate I did it so slowly, that she was rapping out her reply long before I was done. I say all this to my own shame, for Lola must have her due—and I never had a head for arithmetic myself!

When she knew how to calculate time, I put the following question to her: "How many minutes are there in an hour and a half—less thirty minutes?" Answer: "60." "How many hours are there in 240 minutes?" Answer: "4." By this time Lola had also learnt the value of money. About the end of April, 1916, she could distinguish between such coins as 5 Pfennige, 10 Pfennige, 50 Pfennige; 1 Mark, 2 Mark, and 5 Mark, and could compute the value of the Mark in Pfennige. When showing my friends what she could do in the way of arithmetic, her money sums were a special feature and delighted everybody. Here is an example, the date being 31 May: I put the question: "12 Mark less 4 Mark 10 Pfennige?" adding—"Tell me the Mark!" Answer: "7." "And the Pfennige?" "90" (i.e. 7 Mark 90 Pfennige.) Question: "What coins do you know?" Answer: "5, 10, 50; 1, 2." "And what are they all?" "Fenig." (i.e. Lola's mode of spelling Pfennig.) "Lola, how much of a Mark are 50 Pfennige? The answer has to do with fractions." Answer: "1/2." "How much are 225 Pfennige?" "2-1/4." "And 20 Pfennige?" "1/5." "And 60?" "3/5." "And 3/20 Mark, how many Pfennige?" "20." "*No!* "8/20 Mark?" Answer: "15." Towards the close of 1916 I taught her to raise numbers to various powers. At this she was slow in the beginning, but ultimately mastered it fairly well. She could soon answer such questions as—"3^3 = ?" with "27." And—"4^2 = ?" with "16," doing so, moreover, with ease; but up to now I have not been able to take her any further in the matter of extracting roots; in the first place I have had little time to give to it, and secondly, I am by no means on very sure ground there myself! I might, of course, have rubbed up my own rusty arithmetic had my interest in this particular accomplishment of Lola's been greater. But—for my own part, I attach greater importance to the psychological side of this question, and would far rather probe and delve within the depths of her dog-soul, exploring the extent of her other abilities, since arithmetic has already some brilliant exponents in, for instance, Krall's horses.

WORKING WITH OTHER PERSONS.

As may readily be imagined, it is by no means easy to induce an animal to work with any person it does not regard as its accepted teacher. On such occasions, it will behave like a small child, and be restless and even intractable. Often, too, while apparently willing, there may be something unfamiliar in the way in which a question is put (a matter for which no one can be blamed!), this resulting in the impossibility of getting an answer. Sometimes, too, the hand proffered to receive the replies is not held either straight or flat enough, or may not have the right slant that will enable the paw to rap without slipping off. Or, again a hand will be held too high, and thus cause much inconvenience to the animal. Then too, questions are carelessly worded, and seem strange to the method of thought to which its regular instructor has accustomed it, fresh explanations being then required to achieve any results at all. And so it comes, that only those can work successfully with animals who have already been frequently present at the teaching, and are then willing to try their luck, calmly and tranquilly—and quite alone with the animal, so as to carefully develop their own aptitude, as well as gain the confidence of their charge. It is true that in the case of the horses, others, besides Herr Krall, frequently did work with them. Indeed, my father got excellent answers from them, although he had to do with them for only a short time. But the matter seems rather more difficult with dogs; for one thing, they do not stand in front of a board—independently, so to speak—as do the horses; nor are they, from the beginning of their career as habitually accustomed to a variety of persons about them, at least, not to the extent that horses are. And yet they are sometimes quite ready to work with others, this being the case with Lola when I took her to Stuttgart, on a visit to a lady she already knew—Fräulein M. D., and who had put a few questions to her when here at the farm, questions which she had answered quite correctly. At Stuttgart there was a larger circle of listeners, and Lola sat in their midst upon a table. Fräulein M. D. stood beside me, and I asked her to put the question. I do not now remember what the question was, but I had extended my hand for the reply. Lola, however, turned to the speaker, and tapped the correct answer on that lady's arm, giving the second—and equally good one on Fräulein M. D.'s proffered hand. Lola is also in the habit of answering my people with either "yes" or "no" as the case may be, and on one occasion—when I was away from home, having gone to Munich for three weeks—she remained with Frau Kindermann at Hohenheim, and during that time, gave replies to all kind of questions put to her by that lady, as the following report will show:

"REPORT OF FRAU PROFESSOR KINDERMANN IN HOHENHEIM

"On my asking Lola: 'Where is your mistress?' she answered—'minchen!' (München). When I showed her the portrait of my son Karl and asked—'Of whom is this a picture?' Lola at once replied 'Karli.' On 28 October, I received a hamper of vegetables from my mother—known to Lola as 'Mama,' to whom she had been on a visit at Easter. Lola sniffed all the hamper over, then jumped about and wagged her tail joyfully—so I inquired: 'Do you know who the hamper is from?' 'Yes!' 'Then tell me!' 'Mama!' She did a few sums with me every day; told the time; the days of the week, and the temperature. Several acquaintances bore witness to the good work she did—and Lola told them her age—after she had been given the year of her birth. If I happened to be absent minded, Lola knew at once how to deceive me, for she seemed then, instinctively aware that I was not a match for her."

Lola also solved many little sums set her by my friend, Fräulein M. D. (at the time that lady had been staying with me on the farm to gain first-hand experience in the work), and on one occasion when Fräulein M. D. said, "Where is your mistress?" Lola spelt out that I was in the "segenhaus," which was quite true, I having told her shortly before that I was going there. To the great amusement of the maids, Lola sometimes elected to work in the kitchen, with the little seven-year-old son of the housekeeper, and it is reported that her answers were frequently right. I feel sure, in fact, that Lola would work with anyone who was adapted to work with her, and that she would give as good an account of herself, with them, as she does with me.

THE QUESTION OF POSSIBLE INFLUENCE

Eighteenth May, 1916. Lola, who since the middle of April has been accustomed to giving her own independent, and often lengthy, answers, was now rapping very well. Her replies were to the point, decidedly apt, and often quite unexpected. Moreover she usually stuck obstinately to her own way—should I happen to think that something was incorrect, until—on giving in—I sometimes had to acknowledge that she had been right after all. Now, on the 18 May I said to her: "Lola, you must write to my father and thank him for the biscuits, he will then send you some more. This is the way to write a letter, one begins—'dear Father,' or just 'dear,' and then one tells what one is thinking about, you must, therefore, thank him—and when the letter is finished—you must put 'love from Lola'." Now then—begin. Lola started rapping out without further delay, and continued rapidly and "fluently"—so to speak—her letter running as follows: "lib, nach uns kom, ich una ..." (here I interrupted her, believing her about to say "ich und Henny") and asked "is this right?" She said it was: "but, Lola," I urged, "be sure you are careful! ought this not to be a 'd'?" "No!" she said. I was at a loss to make out where this "a" came in, but told her to go on—and Lola rapped: "... artig eben, oft we, kus ich!" So the "una" had been part of "unartig"! (= "dear, come to us, I have just been naughty, often pains, kiss (you) I.") Here she showed that she was quite certain in her own mind, and that in spite of my suggestions as to the form her letter should take, she was yet bent on following her own ideas, since there was no trace of "thanks!" Besides which, instead of concluding with "Lola," as I had proposed her doing, she elected to assert herself by putting *ich* = "I!" "Naughty" referred, probably to a *strafe* she had had about a quarter of an hour earlier for chasing the game, and the "often pain" to headache and to being tired. Anyway, this letter seems a brilliant proof of "independent thinking," and I shall be able to give several more equally fresh and original replies in a later chapter.[19]

Up to this time, it had only been in the matter of *replies* that I had been able to obtain independent communications, but, on 27 May, there was a new development to record: I had avoided asking her any questions for several days, for I had noticed that she seemed extremely tired. But by this day I thought she would probably be fit to do a reasonable amount of work: I have always abstained from this if she showed signs of evident fatigue. So I now asked her: "Lola! how is it you always know when my friend is coming? you knew it before she entered the house this morning!" "Gehört," (= heard) was the reply. "Then, if you know hers — do you know the sounds made by every one?" "No." "Only those whom you know well?" "Yes." Then Lola began wagging her tail near to the door, so I asked: "Who was outside?" Lola gave a "g," and then corrected it with "no." From her delight, I was inclined to think that it had been Frieda, a young girl who had been studying farming with me, and that this was the name Lola was about to rap out. So I discounted the "g" and the "no" and said: "It should be 'f' — shouldn't it?" (note: g = 17, f = 16.) Whereupon Lola continued and rapped — *Frieda*. I then looked out and saw to my astonishment that it was Guste, a new maid who had been in the house about a week. I said to Lola at once: "You were wrong, it was not Frieda, but the new maid — what is her name?" Lola began again — — " ... "and again added "no ..." "Don't you know her name?" I inquired — but Lola replied "yes!" I turned the matter over in my mind, wondering how she had come to rap "Frieda" instead of "Guste," and finally said to her: "Why did you give me a wrong answer, saying Frieda when it was Guste?" and Lola responded with, "You think!" "What?" said I, "did you *feel* what I was thinking?" "Yes." "And do you *always* feel what I think?" "Yes."

This was something quite new, but I explained it to myself, and my view has proved to be correct in all subsequent tests undertaken by me. It is this: *Dogs are susceptible to thought-transference — also, that they are more particularly open to this when tired and when lazy. Further — they are open to such thought-transference even when not actually aware of the question — as for instance, in the present case, where it was a matter of the new servant's name, for here Lola had been able to "tap" my thoughts with respect to what was familiar to her —* (i.e. the name of the other maid) *but* (and this is the most important point) — *a dog cannot receive impressions in respect of matters of which it has no knowledge*!

For example, here Lola could not spell "Guste" in spite of the fact that I was expecting it quite as intently as I had looked for "Frieda" in the first instance; and what is more — I cannot get the dog to "take up" a new thought should she have already "made up her mind" about a matter, as on the occasion when she had been "naughty." It has constantly happened that Lola has held out against me in the matter of some figure in her sums and that—later on—I have found myself to have been at fault, this showing that the numerals "pictured" in my mind can have made no impression on hers; yet, on the other hand, it has also happened that she has accepted my inaccuracies—simply because she was tired, and did not want the trouble of "thinking for herself." Indeed, I could see as much in her eyes—there would be a sense of inertia about her, which indicated that she was only waiting to "guess" by means of *feeling*—a willing receptacle, as it were, ready to receive my thoughts. I have often made the attempt at "thinking" *new* things into her head—but have found this quite impossible.

Shortly after what has here been related, Lola became a "slacker" in the matter of thinking, and kept this up for days. As this pose made it impossible for me to put a serious test, I had recourse for some time to questions only, and—moreover—to questions as to which I could not be sure of the answer, without some trouble or calculation on my own part, for I felt that I might otherwise have really lost my patience with her—unless I had kept on strenuously suggesting the answer—as, for instance: "the stove is green!" Nor did I feel that I could have entirely relied on the inactivity of my subconsciousness, while thus intently thinking. So I kept to such questions as—"What will be the day of the week on such and such a date?" (Naming a date about three weeks ahead.) This precluded any possibility of thought-transference, for I simply ignored reckoning out the days myself. By the way, it is astounding that dogs should be receptive to thought-transference, though there are, of course, many proofs of a dog's acute and delicate susceptibility in relation to the thoughts of human beings, as well as a certain comprehension for a particular situation in which these may be placed. Yet such comprehension can only evince its true force when animals shall have learnt how to give expression to that of which they are aware. With reference to the incident which I have just cited, the thought that presented itself to me first, was that the entire process might possibly be no more than a matter of "suggestion." Yet, on probing further into the question, as well as by drawing comparisons, the conclusions arrived at only further confirmed what I have above stated. That this is so, will, I think, seem absolutely certain to anyone who reads through the whole of this book carefully—indeed, they will arrive at that conclusion without my labouring the question.

It was only by degrees that Lola became amenable to thought-transference, and, in fact, this was only in accordance to the extent to which she became mistress of the human tongue. Now this trait might have degenerated into a serious failing, but, owing to the measures to which I resorted so as to obviate any evil results, it has almost entirely ceased. I now remain quite *passive*, while she is answering, trying to suppress any "thinking *with* her," so that, when she tires, her own individuality may not be disturbed.

ALTERATIONS AND MEMORY

As I have endeavoured to make clear—Lola was, especially during the first month of tuition, exceedingly attentive at her lessons. Indeed, her rapid progress can only be ascribed to this, and to her good memory. Nor did she only evince this alertness at her studies, but noticed everything that went on round about her, even to the following of our conversations, her keenness was surprising. It is probable that every lively and intelligent dog follows what is being said in its presence, and notes our play of feature—this accounting for the demonstrations of sympathy, and other symptoms of partisanship or of aversion they so constantly show. In general, however, such intuitive response is due rather to the dog's memory, and can only be brought to the surface and recognized where the "Spelling Method" has become a familiar mode of expression. Indeed, it may be said that its attentiveness begins then to extend over a far greater field of interest.

On the 19 April, 1916, several ladies—as yet unknown to Lola— were in the room with me. She was sitting near the window and dividing her attention between what was going on outside and in the room. After about half an hour she did some sums and some spelling, acquiting herself very well. For fun she was then asked the name of one of my guests (N.B. the lady's name was really Fräulein Herbster.) (Herbst = autumn, so we usually call her Spring) "What's the name of this girl?" I said: "Frühling" (= Spring) was her reply at once—so that she must most obviously have been listening to us while we were talking.

On the 25 April of the same year, I went on a visit to Hohenheim, taking Lola with me. While there I showed her a picture painted by Ferdinand Leeke and said: "That was done by 'Uncle' who came to stay with us at the farm, at the time when Lola was allowed to go for her first drive in the carriage with the two horses." (This event having made a great impression on her.) "Do you remember 'Uncle's' name?" I added. "Yes!" "What is it?" "leke!" The visit had taken place quite three weeks ago.

On the 20 May I took Lola to tea at S——. She did her work there excellently—both in viva voce arithmetic, as well as in the written tests put to her, and also counted dots, etc. After this the conversation became general, and Lola was not noticed. But in the course of the afternoon I told my friends that I had been to Hagenbeck's Circus a few days before, and that I had seen a monkey dressed as a man, and that it had eaten most daintily, cycled, and done other tricks. This had been a mere casual remark, and in about an hour's time I had returned home with Lola. But that same evening, when I was sitting reading, Lola came and rapped my hands—inquiring— "wer afe?" (= who monkey?) I was at the moment so absent minded that I did not grasp what she was after—but she repeated "afe!" Then it suddenly flashed into my mind—and I did my best to illustrate the performance to her entire satisfaction. I gave an earlier conclusive proof of her memory when I mentioned her recollection of the yard-stick after the very brief explanation I had given her on the subject two months previously. Spontaneous remarks have been allotted a special chapter in this book, and may assist in proving what has already been stated, but I should like here to add an example of how animals put a matter "to themselves," as it were, when the thing *heard* has not been mentally digested, so to speak— or may even be quite incomprehensible to them.

On 26 July, 1916, I said: "Lola! now *you* think of something to ask *me*!" "Yes!" "Well, what is it to be?" "Yes, o h o." "What is the question? What am I to do with that word; the sentence is not complete, is it?" "What means?" "You want to know what *oho* means?" "Yes, yes!"

If we but consider the manner in which a dog will listen—with ears erect—to every word we say, the question Lola put to me will seem most natural! It even "comes naturally" to her to use words which are "above her head," so to speak, as for instance, when she said "surogat"—and in the case of Rolf, who referred to the "Urseele!" (= the primeval soul!) Words such as these are "picked up" by them much in the way that children use words they do not know the meaning of: there may be something in the sound that attracts them, but sometimes they make a guess at the meaning, and in the case of animals, the guess is often a very good one. In Lola this "Art of Guessing" almost led to a sort of Romance!

In my Protocol of 14 December, I have the following entry: Yesterday I asked Lola to tell me why dogs prefer being with human beings rather than with other dogs—and I asked her the same question again to-day. Lola answered: "eid" (= oath). "What is that? you were to answer me to-day: say something properly!" "ich eid." "Oh! I don't understand this! tell me nicely!" "Eid für hunde." "What is *oath* to mean?" "Zu schweigen!" (= to be silent) "*What*? have you promised that to each other?" "Yes." "Who told you that?" "Frechi." (This was one of the dogs on the farm.) "Frechi? and what has that to do with you? Nonsense, had you told me so yesterday I should have known now! Say 'we are happy' otherwise I shall think you are telling me stories: now *why*?" "Wegen iren augen und iren sorgen one ruhe" (= because of their eyes and their sorrows without ceasing). Lola was very tired when she had finished, but it had all been rapped out clearly and carefully, without a single correction. Later I said: "Lola, do you like being with me?" "Yes." "Why?" "ich gut ura?" Now this was quite incomprehensible, so I said: "What do dogs feel when they look at the eyes and see the sorrows of people?" "No." "Yes, tell me?" Then with hesitation: "libe...." (Liebe = love) and to this day I feel touched at these answers. How often in trouble and in sorrow have we not found relief in a dog's sympathy, and been glad to call it a friend in our sufferings? How often has not a dog's eye filled with understanding when its master has sat alone and lost in grief—coming, perhaps, and gently laying its head upon his knees—fixing its faithful gaze on him until at length he might be moved to smile, feeling that—after all—he was not alone? Dogs! may this not be your true vocation? Indeed, this thought possessed me for a long time. This sensitive aspect had not been so apparent to me until now ... I had been so keen on the objective tests and on all that they meant—and now I was almost ready to reproach myself, for had I not centred my love and intelligence on science alone: and only in a secondary sense upon the dog?...

16 December, 1916. On this date I returned to the subject, and said to Lola: "Why do dogs go to people when they see them in sorrow—what is it they then want?" "tresten" (trösten = to console).

"Tell me, Lola, of all the people you know, who has the most sorrows?" "herni ..." But she hesitated, and then turned the "r" into an "n," so that I saw she meant me (Henny)—and yet the spelling had been done with some uncertainty, so I said: "I thought you would have named someone else, whom all dogs love—do you know who I mean?" "Yes."

"Did you mean my friend?" "No." "Who then?" "her zigler!" (Herr Dr. Ziegler) "Then why did you tell a story just now? Did you think I should be pleased to think you meant me?..."

Later in the afternoon Lola was in a state of great depression; "What is the matter?" I asked. "er in or ist aus!" I questioned her more closely, so as to get at the meaning of this enigmatical remark: "What 'in ear'?" (or being meant for Ohr = ear). She replied: "eid zu sagen" (= oath to tell—or to say) adding "ich auch aus" ... (= I also done for). She looked absolutely miserable, and dropped down in a limp heap between rapping out each word, as though bereft of all will-power. I was beginning to feel quite distracted about her: "Lola!" I cried, "Is there no way of putting it right again? Oh, there must be!" "Yes." "Then I will help you!" but again she rapped: "er ist aus!" (Ehre ist aus = honour is gone). She could only answer concerning something she had in her head, and she did so restlessly—though quite distinctly. The whole thing seemed quite incredible! "Lola!" I urged, "how can it be put right?" "e zu...." and here Lola cowered down miserably, and remained so for the rest of the day.

17 December. To-day Lola ran away, returning at length as depressed as ever and bleeding. After I had bathed the wounds on her neck and ears I was glad to find that they were after all, no more than deep scratches. "How did this happen?" I asked. "ich one er." "*How did it happen*? did you run against a tree?" "Dog." "What dog?" "az...." "Tell me properly!" "kuhno." (Kuhno was a fox-terrier in a building near by.) "And were people present?" "Yes." "Who?" "wilhelm." (And this, as I later ascertained, was the case.)

18 December: Lola looked as if she had been crying, so again I said: "What is the matter, Lola?" "No." "Lola! *do* tell me?" "zu rechnen" (= her mode of expression when making evasive remarks). "No, Lola! tell me why you have been crying?" "zu sagen swer" (= schwer: difficult to tell). "No! tell me and I will help you!" I urged (I had incidentally drawn her attention to the above mistake—the "s" instead of the "sch"). "Why difficult?" "wegen er." After a pause I asked again: "Why are you getting so thin, Lola?" (for she had lost flesh considerably during the last three days). "ich so wenig er."

"Wenig essen?" (= you have eaten little?) I suggested—"no"—"Say the last word again." "er!" She kept harping on the same word—Ehre = honour: there could be no further doubt about this, for the missing "h" was of no importance since I had taught her to spell all words according to their sound only—as there would have been no object in teaching her *our* orthography, embodying, as it does, so much that is cumbersome and superfluous.

21 December: Lola was still in the same broken condition: she had been off after the game since about mid-day on the 20th, and had only returned home in the evening. I addressed her with evident displeasure in my voice, saying: "Have you any excuse to make for such behaviour?" "Yes." "Then what is it?" "ich one er." (= I am without honour). "But, Lola! you are only making things worse—if you are naughty and go off like this after the game!" "zu schwer zu leben!" (= too difficult to live!). "Lola! how can honour be made good again?" "wen ich sterbe!" (= if I die!) ... and here the "romance" ended (but not Lola's life!). After a few days she got better and soon became as lively as ever—the wild and excitable creature she is by nature, whom none would take to be the mother of four children—and a "learned dog"—into the bargain! The thing is—could the dog have caught up an *impression* from some human mind—something she had heard said in conversation, and which she had—in some mysterious way—assimilated and applied to her own life? I cannot tell, but I almost feel as if this must have been the case. There can be no doubt that animals *have* a sense of honour, yet it would seem unlikely for it to function in the manner above narrated. Yet how much remains still unaccounted for within a dog's soul—how many attempts at unravelling will have to be made before the right clues have been touched, which shall lead us to our goal within this labyrinth. There is so much which it is impossible to bring into co-ordination with the human psyche, for though there are many fundamental impulses, common to both man and beast, we cannot approach the subject, nor yet measure it according to our human standards, where the psychology of a dog is in question. Another thing: in educating these dogs specially reared for experimental work—we should be careful on no account to suppress those instincts, which are natural to them as *dogs*—i.e. their "dog-individuality," transforming this—either by praise or blame. Just as

certain conceptions and feelings, held by different peoples differ fundamentally, so too, has every animal a *something* which is *its very own*, an *innate something*, and this—in order to successfully accomplish our ends—must be held inviolate. Now, this is, of course, very difficult—since to instruct and educate an animal is, of itself, an infringement on its true nature—and, indeed, the same might be said respecting the life it leads among human beings. Yet I believe that where an animal *feels* that its own inner nature is left unmolested we may often succeed in "*hearing the animal speak within the animal*" (if I may so put it), rather than its "human connexion." That sentence of Lola's: "wegen ihren Augen und Sorgen ohne Ruhe" (= because of their eyes and their sorrows without ceasing) certainly "rang true"—one could feel it as the answer was being given—yet—where the meaning is dubious, as in some of her replies which followed this one, decision becomes difficult indeed!

THE CONNEXION OF IDEAS

The ability to definitely connect one idea with another is clearly apparent in the animal mind, and may be attributed to its excellent memory and powers of attention. In everyday-life this becomes apparent as the reflex of their experiences, the impressions of which, having once impinged on their sensibility have left their mark, so to speak, and this experience thus practically acquired, shows itself at times as the shrewdest of wisdom, even though we may now know how their "power of reasoning" was arrived at — without words. We need only think of the way in which animals have time and again rescued their masters — going for assistance in the most intelligent way — this being but one of the many examples which occur to my mind. Nevertheless, a combination of thoughts, such as is carried out purely on the *mental* plane is only possible in the case of an animal that has been trained. I had a very pretty example of this on 14 September, 1916. I had taken Lola with me to a neighbouring estate. The rain was coming down in torrents, and we sat beneath the sheltering roof of the balcony and gazed out at this flood. "Where does the rain come from — Lola?" I asked; "uzu," she replied. "And what does that mean?" I queried. "heaven." "And what is the water wanted for?" She hesitated and tapped — "ich zu taun!" "What does *taun* mean? tell me differently!" (as I thought she was evading a direct answer). "funo!" "Nonsense!" "yes!" "I want to know what *taun* means!" "when I don't hear!" "Nonsense! *'when you don't hear!'* — there is some letter wrong!" "yes." "What should it be?" "b." "Taub?" (= deaf). "yes."

A week earlier I had explained "eyes" and "ears" to her, and the meaning of blindness and deafness, and yet could not make out why she was now using the word "taub" in this connexion.

"Did you mean that you did not understand me?" "no." "Then why did you say that?" "ich er (rather reluctantly) ... or ..." "Well — — ? and what more?" "I won't say!" "You won't tell me?" "yes!" The next day I returned to this question, for I could not make out why she gave me such answers, and made such excuses. She well knew how determined I could be in the matter of "catechising," and that I will stand no "nonsense" when she begins her little game of rapping "1!" — the meaning of which, she had once informed me, was *"I won't tell!*" and the sequel to which I generally found to be that she would put me off with any word that might just happen to come into her head. But why had this remark occurred to her yesterday? I wanted to get to the bottom of it, so returning to the attack, said: "Why wouldn't you tell me yesterday what water is good for?" "I thought of ear!" "What has water to do with 'ear'?" "water in ear horrid!" Here, then, was the reason! In her very fear she had not been able to bring forth her true answer—for, owing to me, the water had got into her ears—and made this lasting and unpleasant impression—when she was being bathed—or when I threw her into a stream! The reader may already have noticed other instances where a direct connexion of ideas has occurred. I have purposely abstained from pointing to the obvious in each case, believing that anyone who is keenly interested will do so quickly enough for himself, and I am loth to weary my Public by needless repetitions.

SPONTANEOUS REPLIES

Spontaneous replies provide a special proof of this ability to form independent thoughts, and is found both among horses and dogs. Such a reply is indeed the sudden and evident utterance of some thought, and of a thought which—to it—transcends all other thoughts at the moment: one which regardless of all other questions which may at the time be put to it, looms largest, and the animal will therefore utter this remark, asked or unasked—and quite independently of any question, but more after the manner of "making an observation." Such a thought may have nothing to do with the subject in hand, and persons who are participating in this conversation *à deux*, can only arrive at the inference of ideas after having carefully thought the matter over—it may also be that they will fail to see any association of ideas at all. Now, it is indisputable that such replies belong to the most important category—for they may serve as proofs to those who themselves have not worked with animals for any length of time, and who, therefore, cannot become sincerely convinced as to the truth of the matter by travelling the longer road of personal test and experience. The teacher of any horse or dog of good parts does not need this proof: there are thousands of small instances which in their sum total prove important—trivial and uncertain though each one may be, when regarded by itself. It would be difficult to know how to convey these to anyone in words: glances, movements, a certain "live appeal"—it would require a poet to catch and fix—in short—to idealize—telling us the true inwardness, so that we might indeed comprehend ... and even then he would, I fear, make for weariness, when grappling with what well may seem interminable.[20] Here are a few examples:

16 May, 1916: Lola was doing arithmetic and I had given her some new sums. Suddenly, instead of calculating, she gives—"not reckon." I asked her the date, she replied "16"—adding of herself "too little to eat." In the course of the afternoon, Lola, who had gone with me to tea at B. L.'s, was shown some pictures: "What is that?" she was asked. "re," (ein Reh = a deer) "segen haus, ich wenig nur arbeite." "Will you do more here?" "yes." "Arithmetic?" "Yes, yes!" (very joyfully) and excellent replies followed.

3 January, 1916: On this date I began teaching her the capital letters of the Latin alphabet; A = a, B = b, and so on, when she suddenly "butted in" with "go out." As she had worked very well up to that moment I opened the door and let her out. But in five minutes she was back, looking anything but pleased; "Well, didn't you like it?" I asked; "no!" "Why?" "come too!" I venture to think that I have here given good proof in the matter of "spontaneous" utterances, the best, perhaps, being the one given at B. L.'s, where she complained of having done insufficient work, for her fault-finding was generally the other way round! But she has always loved to show off in that particular circle, sensing no doubt the friendly interest taken in her there.

WRONG AND UNCERTAIN ANSWERS

If Lola is tired she will either not work at all, or—at most—work badly, which is but natural! Yet there is another and even more frequent reason than fatigue for her indifferent work. The dog may to all appearances be bright and fresh—leading me to expect the very best results, and yet—with everything seemingly in her favour, she may that day be an utter failure. This is particularly unpleasant if on one of these occasions visitors happen to be present, and more especially should there be sceptics among them. For this failure to respond where the subject happens to be one in which she has repeatedly given brilliant proofs of what she really *can* do, is embarrassing and humiliating, for then those who are only too ready to scoff merely feel their case strengthened. Indeed, it needs some determination to keep one's temper on such occasions, yet to "let oneself go" even for one moment—would mean weeks of painful and laborious uphill work in order to regain the dog's confidence. One is often entirely at a loss as to the reason of this "inward withstanding," which may even elude long and careful investigation. Now and again the answers may not be forthcoming when one is alone with her, and behold—! a stranger enters the room, and she becomes all friendly eagerness to do her best: then again, the exact reverse of this may be the case, or on some days she may be useless both alone and before company. There have been times when she has been delightful and engaging in every way—till work was mentioned ... when the whole expression of her face would change, and she would assume her "stupid look," deliberately, so it would seem, rapping out the simplest answer wrongly! The very act of rapping is at such times a mere careless dragging of her paw—as though it had nothing to do with the rest of her body. Pleading, threats, the nicest of tit-bits—all are then unavailing, and she remains *seemingly* idiotic—the mere sight of her being enough to drive one wild!—for low be it spoken—*it is the sheerest impudence*!!! Indeed, the visitor who does not know her, and happens to "strike" on one of these bad days, would have to be dowered with more than his share of amiability and imagination, should he be able to mentally visualize anything approaching "brilliant accomplishments" in the face of one of these fiascos. Whether these "turns" be due to sudden obstinacy, to some feeling of injury inflicted either by myself or the onlooker—to what on earth such tempers be due I cannot tell! but I have put up with this sort of thing for two hours at a stretch sometimes, keeping my self-control till at length I have had to rush out of the room—relinquishing every hope of victory for that day, and with a feeling of what seemed almost hatred against this unreasonable beast! although I must say that such feelings do not last very long—for I am not a good "hater"—and then ... Lola would soon try to "make it up again" in some touching way!

I may say that for the first four months she worked splendidly before strangers, and quite as well with me, but from that time onward her work was equally *uncertain*—both in the presence of others and when alone with me. I know of no cause for this, I can only say that I often seemed to "sense" about her a feeling as though she considered these labours superfluous; as though she had become in a manner "disillusioned" as to the "results" accruing from her work. Was the praise, or were the rewards inadequate? the fact remains, that on such days utterly senseless answers were the most one could get after constant and persuasive questioning, while the solutions of her sums would be completely wrong. When once the novelty was gone, indifference and lack of interest soon took its place, and this applies to everything she learnt. In the beginning, close attention, and keen alertness—resulting in ready and intelligent replies, then a sudden slackening, so that it would seem useless for me to pursue the same subject again for weeks. This tiresome trait (which, by the way, I can in part appreciate) may, I fear, in time attack her spelling too—and then everything will be over, as far as Lola is concerned. Not that she will be getting more stupid with increasing age! indeed, as she grows older, she will probably be better than ever able to understand what is said to her, but she will no longer find it worth her while to pull herself together so as to do decent work. I shall, of course, do all I can as far as trying to influence her so as to put off the evil moment—but the fact is that one has here to do with a remarkably sensitive and obstinate living-creature, and one that is quite able—though in a passive way—to maintain its own standpoint.

I shall now give a few specimens of the *almost* unintelligible answers dragged from her, as it were, after much grave reproach:

16 August, 1916: "Lola, rap something!" "mal one lif unartig sein." "What is the meaning of 'lif'? do you mean 'when you ran'?" (lief = ran, the past tense of laufen = to run). "no." "Did you learn that word from me?" "yes." "Then explain yourself." "ich rante in wald zu re" (= I ran in the wood after deer). Apparently she was in no mood for explanations, and it was only after wrestling with her that I could get any sequence of words at all. At other times when urged to get on with the subject she will in her contrariness rap as follows: "o zu ich" or "e wo zu" or "zum zu wozu" or "we" and so on—letters with which it is rarely possible to put together even such small words as *wo* (= where) or *zu* (= to, or for) and the longer one persists on such occasions, the more senseless her remarks become; it is the rarest thing for her to suddenly pull herself together so as to give a proper answer. And here again I can find no excuse for her behaviour; though it may be that she dislikes my persistence, and therefore has recourse to any nonsense by way of a quick reply! So as to get her in some manner to recognize the errors of her ways I have again and again persevered with the utmost patience, so as to arrive at some consistent answer—yet all I have succeeded in arousing, has been increased reluctance on the dog's part.

MATTERS WHICH — SO FAR — ARE UNACCOUNTED FOR, OR UNEXPLAINED

As will, indeed, be evident, there is still much that remains unexplained; much that it will be the task of the future to throw light upon. Tests which have been but uncertain in their results; accidental discoveries, the importance of which only becomes evident, after the results have been tested in connexion with a number of animals. Among these may be placed the more recent experiments dealing with the sense of scent, undertaken by Professor Jaeger, and in this category should be placed also what I think to be a rather interesting test connected with Lola: I was at the time staying with my family at Hohenheim, and I asked the dog how many pups her mother had had — including herself: she answered "12." I inquired of Professor Kraemer if this was so, and he said that at the time at which he had seen them there had only been eleven. I then made the same inquiry in Mannheim, and found that there had been twelve, but that one had died immediately after birth. It was the only instance of which Lola knew about a dog having pups, so one day I asked her in fun (19 June, 1916). "How many children will you have?" (Thinking that the answer would be 12). At first she replied with "yes!" "Do you know how many? why that's impossible!" But she rapped "9." "How many boys?" I asked. "3." "And how many girls?" "6." I thought that this statement was due merely to her desire to make some answer, so I put the same question the next day — but the reply was again, "9." So I told my friend about this and we awaited the interesting event in much suspense — it took place on the 22 June, 1916, in the presence of my friend, the housekeeper and myself and — *there were nine puppies!* two males and seven little lady-dogs. I kept two of each, the others being put to death at once by one of the farm hands, for — owing to the war, as well as to the fact that the pups were not thoroughbreds, I could not undertake to bring them all up. But, the question is — how could Lola have known that there would be nine?[21]

ALTERATIONS IN CHARACTER

As a result of all that has here been stated, the question may very naturally arise: are there any indications such as lead to suspect a change of character, or do any other practical results follow on these educational tests? Now, Lola is by nature lovable, lively, full of fun, and she has retained these traits to the present day. Her great excitability has diminished, it is true, but this is probably due to her having grown more staid with years. Yet a difference is also to be found where her character—her dog-soul—is in question: it may be noticed in the suspicious way in which she now regards people, as though she were "drawing comparisons" between them and herself. We have, in fact, fallen somewhat in her estimation. She "asks"—so to speak—as to where our vaunted superiority may lie, and would seem to compare her newly-acquired knowledge—together with the existence forced upon her—with the life that is ours. Since she has made these "educational advances" one can often see in her eyes something that amounts to an angry reproach—something like an impatient question, as to *why* we have so much food and freedom as compared with what is meted out to her. She follows our thoughts to a great extent, and our abilities no longer seem to impress her, since—to her—it is only those which she herself has mastered that come under this heading at all, and here—a slight contempt for the "oppressor" is often discernable. There is also a greater show of independence and frequent contrariness, owing to her diminished respect for our "species," in short—it becomes more difficult to deal with the dog. The days of blind confidence are past—even though an innate sense of devotion to man remains, for what has just been said, seems always to occur more as the result of "moments of reflection." Indeed, this entire educational process would have little that is joyful about it, were it not for the feeling that the animal understands its friend, and is in a position to converse with us within certain limits, and this outweighs and compensates for all the rest!

As to the practical results—I can say little that is favourable. The dog's *thinking* seems to be at variance with her acts: thought can therefore, have little influence upon a dog's behaviour, for—as has been the case with dogs of every kind, from time immemorial—its actions are due to the excitement of the outer senses, such as scent, taste, and hearing, and any emotions observable are but the direct and inward continuation of those external sensations, and, as such, last but for a given time. What we may term the "thought form" that is bound to any given *word*, representing objective thought in its simplest form, rotates within a very limited circle, and is powerless over the animal's feeling. For instance: Lola knows that she is forbidden to "hunt" i.e. to go after the game, etc., indeed she has shown in many of her replies that she is well aware of what "totgeschossen" (= to be shot dead) means. And yet—once the scent is up, off she goes, and nothing will prevent her—for, she *must* go!

This is a particularly strong characteristic which beating and being deprived of her food may sometimes *check*, but which her own powers of reflection do not cure: and it is the same thing with most of her faults. At times it will be unreasoning obstinacy, but even where she uses a certain amount of reflection, the *result* is identical. It has been no better where—with the help of thought—we have endeavoured to bring about actual results. An animal can be got to understand and carry out certain injunctions, such as—"sit up and beg," "lift up your paw," "go to your bed," "go out of the door," and much more of the same description, while after instruction it will understand "behind the stove lies a biscuit," yet *action* seldom results from such knowledge. The dog's eyes will brighten, and it is evident that it has perfectly well comprehended the meaning of the words, indeed—this much can be easily ascertained by questioning it—but the dog will seem incapable of translating what it has comprehended into action. At such times Lola will rush about, as if her limbs would not obey—as though the influence she could bring to bear on them was not sufficiently powerful—and the final result is excitement. Connexion with the motor-nerves does not come into being in response to the action of the cerebrum. As the result of repeated written and spoken orders it is possible (with a certain amount of additional aid) to set up this connexion from without, yet, even then, the actual effect is but moderately successful. On the other hand, action in the reverse way—from the nerves or senses to the brain—is easy where the dog is concerned. Lola can report about things she has done, such as—"saw deer," "drank milk," "went into wood," "was naughty," "ate some of the cow," for reflection gives more time to master the subject, and to notice what is past, and this will therefore show, that in the way of practical results, the best will be those obtained by asking a dog what he has seen, heard, or scented, etc. Indeed, it is along these lines that the police dogs have proved their worth and importance. Yet it is very necessary that one should make sure that one's dog is not a liar, but an animal capable of taking up its job in the right manner. With our present knowledge, however, we are unlikely to achieve very much, since we cannot say to a dog—"go here or there"—or—"take this letter to so and so."

Not but what dogs have—in exceptional cases and after training—learnt to carry out such instructions, but it has resulted *without their thought-activity having been developed*. They get familiar with a certain road, and—basket in mouth—they will proceed to the baker's but—independently of habit and external impression—by the mere appeal to the brain or by means of the most persuasive words, we can attain to nothing worth mentioning, nothing that could be of distinct value, where a dog is kept for use. The sense, the object, and the reason for this educational work must be sought on other grounds.

A VARIETY OF ANSWERS

It was some time after Lola had mastered the art of spelling before I was able to get her to make independent replies. The first of these was given on the 13 April, 1916, and from that time onward they became easier and more frequent: most of those I have set down date from that period. These answers were at once noted, according to their numerals, and when the sentence was complete it was transposed into letters of the alphabet. Whenever there were any spelling mistakes, the words were placed before her, and she was told to name each successive wrong letter in reading over her answer. As *I* knew the equivalent letters, I was able to write them down at once, and if the reply was a short one and no paper at hand, I could memorize the letters, and enter them in a book as soon as the lesson was over—adding the questions to which such answers had been given as well as the dates. All other questions and answers, as well as particulars relating to new exercises were also set down here.

Here is an answer I received from her on the 13 April, 1916: Lola was staying with me at Hohenheim, where we had arrived on the previous day, and I proceeded to Stuttgart in the morning. When I got home in the evening I asked Lola: "Is it nice here? have you had good food at father's?" to which the answer—quite wide of the mark—was—"wo wald?" (= where is the wood?) For I had been telling her about all she would be able to enjoy and that, among other delights, there would be the woods; as however, her afternoon walk had only lain through the fields, her mind was now absorbed with the one idea—"where was the wood?"—to the oblivion of everything else.

15 April: On this day the written question was put to her: "Why does Lola like going in the woods?" the reply was at once forthcoming, and I dictated it to Frau Professor Kindermann. "Where there is wood also deer and hare"—she was not quite clear in her spelling at first, indeed, in this respect she sometimes reminds one of a foreigner—as also in the matter of her grammatical mistakes.

The next day, after having done a few sums to please some friends who were present, she was asked: "Who is the dog in the room?" "I!" she replied—not "Lola" as we had all expected. (Rolf has as yet never alluded to himself as "I"!)

Two days later she was asked in writing: "How many dogs can reckon and spell?" To this she began her reply in a very brisk and lively mood, but soon wavered, as though at a loss for the right expressions, then followed a short pause—and finally she resumed her rapping with renewed animation. The reply, it will be noticed, is detailed, and does not keep to the plain question that had been put. "how many have been taken (for it)? Rolf talks, counts; two more" (short pause) "I also, also heinz and ilse." For, so as to fire her ambition, I had told her about her brother and sister, Heinz and Ilse.

19 April: "Lola," I asked, "what was it that ran away from you on the meadow?" "cat!" "What did you want to do with the poor cat?" "kill!" "Have you no pity?" "no!" "Then is the cat right if she kills you?" "*no!*" "Why?" (The reply to this was rapped indistinctly.) "Have you no pity for any man or animal?" "for dog!"...

22 April: I had told her that my brother was coming, and that he wore a field-grey coat and was a soldier. When he arrived I said to her: "Who is this?" "Your brother."

Next day she was asked in writing: "What did Lola see swimming in the water?" "duck!" I had shown her a duck on the previous afternoon.

26 April: On this day Lola appeared before Professors Kraemer, Mack, Kindermann and Ziegler, of Hohenheim, which resulted in these gentlemen forwarding the following statement to the "Mitteilungen für Tierpsychologie" (= Communications respecting the psychology of Animals), series 1916; Number 1, p. 11:

"EXAMINATION OF LOLA BY PROFESSORS KRÄMER, MACK, KINDERMANN AND ZIEGLER

"In our presence Lola solved a number of sums, such as: $5 + 8 = 13$. $30 + 10 - 15 = 25$. 4 Mark - 1 mark 20 = 2 mark. 80.

"She next counted the number of persons present. After this, several dots were scattered about a sheet of paper: at first she put their number down as 19—but corrected this to 18. Lola then told us the time: it was 4.16m., and after this she did some spelling. When shown the picture of a flower she rapped: "blum" (Blume = flower), and to my somewhat faulty drawing of a cat she responded with "tir" (Tier = animal), while finally to the question of what was the name of the Mannheim dog she replied "mein fadr" (Vater = father) — we all having expected her to say Rolf. Then followed the musical tests which amazed us most of all, for here she exhibited an ability lacking in many an individual."

27 April: Lola very tired: groans and does everything wrong. I said: "Are you lazy?" She replies "no." "Then why are you answering so badly?" "go!" "Who is to go?" "*tired!*"

29 April: I asked Lola why she had not attended to me on the 22nd, when — on a country expedition we had made together — she had insisted on running after the game when I had called her back. I had had to hunt after her for ten hours the next day, finding her — by the merest chance — at a peasant's house. She had settled down there alongside of a sheep-dog to watch the sheep, and seemed by no means pleased to see me; usually she is delighted! Her reply on this occasion was — "Lola went in wood, also lay down and was hungry." I returned to the question later in the afternoon when she made the rejoinder — "sought, didn't find."

30 April: Once more I returned to the incident mentioned above and Lola answered "to marry a dog" — (the consequences of this escapade becoming apparent, when Lola presented us with her litter of pups on 22 June). Then Lola added a spontaneous remark on her own account for, seeing a biscuit in my hand, she rapped "I to eat!"

On 1 May little was forthcoming in the matter of arithmetic — with which we always began our lessons, for Lola rapped: "too tired."

3 May: In reply to my question as to what she had had to eat at the peasant's house she said: "milk."

The next day I asked her "where is my friend living now?" to which she answered. "Hanhof." (N.B. A name under which she includes the entire district). "What is the colour of the woods now?" And she answered. "Green." Then "Why are you looking at me so crossly?" "We." "In your head?" "Yes." "What has given you a headache?" "Learning."

8 May: Lola had been rolling herself about in some frightfully smelly mess—a thing she, like other dogs, never loses an opportunity of doing. "Do you *like* that smell?" I asked. "Yes!" "But don't you know quite well that I do *not* like it?" "Yes!" "Then why do you always do it again and again?" "I love it so!" The same afternoon, after her musical tests, the maid came into the room to lay the fire. "What is Kätchen doing at the stove?" I asked. "Fire," replied Lola.

The next day: "Lola! who do you like best of all people and animals?" "Ich!" (1). "If you mean *yourself* you should say "mich" (myself)", so she at once rapped "mich!" "And after yourself?" "Dich!" ("thee," the familiar of you commonly used in German). A frank remark, at all events, and without the taint of human egoism!

10 May: Lola has been gnawing a bone: not knowing of what animal it was, I put the question to her and she replied: "re" (reh = deer). The truth of this being confirmed in the kitchen. I then asked: "What bones do you like best—deer, hares, wuzl" (this is her own name for a pig), "or ox?" Answer: "Wuzl!" "Are you pleased that you know more than other dogs?" "No." And then—as though after due reflection—"no!" (*Emphatically.*)

11 May: I showed Lola a biscuit, shaped rather imperfectly in the form of a fish, saying: "What is this—an animal that swims in the water?" Reply: "Fish!" In this case I do not think she had really recognized it, but had named the only animal she knew of connected with water, which—after all—was rather clever of her!

12 May: "Lola!" I asked, "would you like to be a human being?" "No." "Why not?" I asked—showing her a biscuit. She (promptly): "I eat!" "No! not till you have answered!" "Because of work!" A little later I said: "Do you belong to me Lola?" Very energetically—"No!" "To whom do you belong then?" "To myself." "And to whom do I belong? do I belong to you?" "No!" "Whose Henny am I?" "Your own!" These amusing answers bear the very impress of the animal's sense of independence: she is loth to be considered a "chattel," like some chair or table!

17 May: In the presence of my friend and of two dogs I asked her—"Lola, why don't you like Dick?" (Dick being one of the dogs present.) "Too wild!" was Lola's comment. "What do you like best to eat?" "Ich ese wi so mag!" "Is that quite correct?" "No." "Which word

should be different?" "4!" "Then what should it be?" "Ich." "So it is to be: ich esse wie ich mag?" "Ja!" (= I eat as (or what) I choose.)

31 May: Lola did her sums badly, and I spoke very seriously to her; after which she improved, rapping out an independent remark: "say I am good!" She wanted to hear that I was ready to "make it up" again! That evening, some friends being present—I wrote on a scrap of paper—"bon jour!" showed it to her for a moment and then removed it, saying: "now rap what you have read!" And she rapped: "bon jur!" Having only missed out the "o"; the word had not been spoken, so that I had naturally thought to see the "o" among the other letters.

2 June: Lola was to write a letter to a lady whose daughter had been staying with me on a visit. The dog was much attached to this young lady, and had frequently worked with her. She began her letter with all sorts of nonsense so that at length I said: "First rap 'dear' and then tell her about the biscuits you had from Irene."

The letter: "Dear, certainly Irene is very nice to me" ... then "were" ... "What's the meaning of that?" I interrupted, but Lola lay down and said "Zu we!" (= too indisposed.)

3 June: "Will you work now?" "No—we!" "Where have you a pain?" "O sag!" "What am I to say?" "Oh seh!" "But what am I to see?" "Ich!" "I am to look and see where you have a pain?" "Yes, yes!" But these "pains" seemed to have been called forth by laziness and possibly some slight fatigue.

15 June: A lady has come to stay with me for a few days and I said to Lola: "Why do you like Fräulien Grethe?" "Ich is zu artig." (This is indistinct but probably meant she is kind to me.) Presumably she could think of nothing else to say.

25 June: Lola had been brought indoors—away from her young family, and I said: "Is there anything you would like to have in the stable, now think?" "wenig uzi!" "What is uzi? do you mean music?" Answer. "Lid" (= lied.) "What is that—singing?" "Yes!" "Do you like to listen to us when we sing?" "Yes, yes!"

24 July: "Lola! now think of something I am to do: give me an order!" (By the way, in reply to a similar question put to Rolf by the wife of Colonel Schweizerbarth, at Degerloch, he had commanded

her to "wedeln" (= to wag!) N.B. This word being only used in connexion with *a tail* in German!) But Lola merely ordered me "to work" — "What am I to work at?" I inquired. "Raking the garden, reckoning, writing or reading?" And I was somewhat surprised, for she was used to seeing me at work at something or other for the greater part of the day; but after mature reflection she added — "ales" (Alles = everything).

27 July: To-day I invited her to tell me something she might be thinking about, adding: "Will you say something?" "Ja, esen." "Oh, Lola!" I said in desperation, "why all this talk about eating! about food! don't I hear enough of it from senseless labourers and maids? and now you begin too! It can't be otherwise, at present: say something else!" "Ich am esen" ... "What? *again!* well go on" "... zu wenig narung." "Ich am essen zu wenig nahrung" (= I from my food (derive) too little nourishment). "Ja!" Poor Lola!

10 August: To-day is my father's birthday: he is staying with us, and Lola was to give him a "good wish." I suggested all kinds of things, such as good health; long life; and so on, but she would have none of them. At last she rapped "Ich wunsche esen"; and after a short pause she continued, "... und ich auch" (= I wish him food and for myself too.) "Now give him a second wish: something you yourself find good." So she said: "Re jagen und has...." "And a third?" "Heiraten" (= to marry). Such were the dog's wishes for my father's natal day! Food, Hunting and Marriage ... the first one being ever the central idea in a dog's thoughts — and yet, how necessary are all these three wishes to the maintenance of species — "urged ever onward by the driving-power of hunger and of love!" after all — there is something very simple and direct about an animal!

30 August: To-day I asked Lola: "Do you wish every one to marry and have children?" "No." "Why not?" "Arbeiten unmöglich," (= work impossible). "Go on: if it is impossible, one simply does not work!" "Und ausgen ..." "Go on?" "Auch zu vil esen!" (und ausser dem, zu viel essen = and besides that, too much eating). Here spoke experience.

1 September: Lola was shown some dots on a sheet of white paper, but declined to count them. "Why won't you count?" "Ich ursache one wisen!" (= I have a cause (reason), without knowing (it)).

Then she began to tremble violently, and I asked her why—to which she replied: "Ich kalt" (I (am) cold).

2 September: An old farm labourer and his wife had come to my room to see the dog, and in their honour Lola consented to do some sums. The old man was delighted when, on my suggestion, Lola spelt out his name: she rapped "Wilem," and when I said: "Did you hear that from me?" she answered: "No." "From his wife?" "Yes!" This accounted for the spelling, as the woman is from the Rheinland district, and says "Willem" for Wilhelm.

6 September: "Lola, why did you bite Jenny, yesterday?" (Jenny is a terrier lady-dog.) Answer. "Em ..." "What does that mean?" "Wüst a—a und renen." (= she was a dirty dog and also hunted.)

7 September: Lola came in from the farm quite wet, and I wanted to know the reason of this, as only the woods were still wet from the recent rains. To my question she made answer: "I in wet." "Were you in the grass or in the woods?" I demanded. "Grass!" "Is the wet grass nice?" "Saw deer in wood—why I came to you!" In spite of such a tempting sight, she was evidently in a virtuous frame of mind: in earlier days she could never resist giving chase.

8 September: "Why are you not eating your food? is it bad?" "Yes!" "What is wrong about it?" "Smell!"

20 September: "Lola," I said, "give me the reason for why you are alive! do you know one?" "Yes, no."

The next day: "Now tell me your answer as to why you are living?" "Yes!" "Well?" "Egal ich lebe gern!"... (i.e. *egal* is an expression of indifference, such as "*it is all the same to me*, I like living"). How simple and complete is the dog-point-of-view! "And is that all? didn't you wish to add something more?"... "in Welt" (= in (the) world). The expression "egal" she will probably have picked up from me.

22 September: To-day I noticed by Lola's behaviour that she wanted to say something, so I put the question to her, and she replied. "Yes." "Well, go ahead!" "I wish to pay you for getting food for me!" "Do you want to give me money?" "Yes!" "But, where are you going to get it from—can you tell me that?" "Yes!" "From where?" "From you!" There was something quite logical about this way of arguing, for Lola had heard much talk about money, farm-hands

being often paid by hour—and she had no doubt been an attentive listener and observer, at such transactions. Then—all of a sudden—she rapped. "I without work!" "What do you want to have?" "Haue!" (= a beating!). I thought I had misunderstood her, so repeated—"haue?" "Yes!" "Say something else!" "Reckoning." But the fact remained that she really longed for a beating—not having had one for a long time, for to my repeated inquiries she kept on with "Yes!" So at length to make sure, I fetched my riding-whip and gave her a light flick, saying—"Is that what you want?" "Yes!" "And do you want more?" "Yes!" she insisted, though all of a tremble, and—unwillingly enough—I had to administer one more.

13 November: Lola had to write a letter to a lady of whom she is very fond: it ran as follows—"dear, I have just been in the yard, I like eating biscuits, I kiss you!" (I think this letter bears the evidence of being Lola's own composition!) Later in the afternoon, when she was out with me, I saw a notice put up saying: "Dogs are to be led on a leash"—and I invited her to read it, but she would only give it a glance. Both on our way back, and when we got home I returned to the subject, saying: "What was on that notice-board?" But she rapped "No!" "What? you mean to say you don't know?" She had, however, already started rapping again—"ich unaro...." "Go on! surely the *o* should be a *t*?" (Thinking she meant unartig = naughty). "No!" "Then what should it be?" "No." "Is it a dog's word?" "Yes!" "Well, tell me in a way that I can understand!" "No!" "You can't do so?" "No!" "Say something like it!" "Ja! ich irre, ich es ansehe morgen!" (= yes! I erred, I (will) look at it to-morrow!)

On one occasion I had explained to her that there were also other languages; English and French, for instance, and I now once more tried to influence her memory by my own thoughts.

"Lola," I said, "do you know what is meant when I say—*je veux manger*—do you understand that?" "Yes!" "Then tell me!" "Ich wil esen!" "But do you understand this: *il faut que je travaille*?" "No!" "Think again!" "No!" "Travailler?" "No!" This proving that what I had not taught and explained to her she was incapable of saying—or rather, spelling.

15 November: The following incident was communicated to the "Mitteilungen of the Society for Animal Psychology" (series 1916, No. 2, page 74), by Professor Ziegler:

"Lola had been for a walk with Professor Kindermann, and on her return was discovered to have a feather in her mouth. Fraulein Kindermann asked her: "What animal's feather is that?" she answered: "Hen." "How did you come by the feather?" "Killed hen!" "Why?" "Eat up!" "And have you eaten it up?" "No!" "Why did you run away?" "Fear." "Of what were you frightened, of people?" "No!" "Then of what?" "Ursache!" (= cause, i.e. cause of fear.) There is something rather charming here in the way in which the dog confesses to her misdeeds, and at the same time owns up to having a bad conscience!"

16 November: Lola must have noticed to-day that there was roast hare on the midday dinner table, for in the afternoon when invited to make some remark she rapped: "Zu wenig ..." (then hesitatingly) "h ..." "Are you afraid?" I inquired. "Yes." "Nonsense, I shall not scold you!" "... as!" — "Zu wenig has — who?" (= too little hare) "Ich, o we!" (= I, oh alas!)

18 November: To-day she started to rap nothing but nonsense; but in time it became more distinct, and ended up with "ich zälen!" (= I (wish to) count). I asked her if this was a fact — and she promptly said "No!" She then kept on making her usual sign that she wanted to go down into the yard, so I let her out, but soon she ran up again quite briskly, and at once rapped out clearly and distinctly. — "Warum ich und sie so rau geartet?" "Is this what you mean?" "Yes!" "And — who is si meant for?" "Heni!" "*What?*" I exclaimed, "you are suddenly addressing me as *sie*?!"[22] "Yes!" "But Lola! that is what we only say to people we don't know well! you have always called me *du* because you were fond of me — isn't that so? are you saying *sie intentionally* now?" "Yes!" "Yes? but why?" "Because strange!" "How *strange*?" "Yes!" "Was: warum ich und sie so rau reartet (= why are I and you so roughly constituted?) the end of the sentence you began before?" "No." N.B. In this manner did she wish to lodge her complaint, so to speak, against me for not always understanding her when she prefers to try and "rub in" the meaning of her faulty spelling, by gazing at me in her "intent" fashion — indeed, I had

always sensed her annoyance at times when she had not been able to gain her ends in this way! In simple matters, such as "wish to eat," or "go out," I could of course, guess her desires, but she was of opinion that I ought to be more "understanding" still—and this is difficult!

1 December: "Lola, what will become of you when you are dead? what will become of your body?" "If..." "No; that is no answer! You are to spell properly!" "Zu esen für wurm" (= food for worm.) "And, Lola ... your soul? do you know what that is?" "Ja, nur get in himmel!" (= yes (it) only goes to heaven!) "Did you hear people say that?" "Yes!" From this it would seem that any seeking after the dog's own sensations on the subject are useless. By the way, some time before I had read Rolf's remark to her: "All tier hat seel, guck in aug" (= all animals have souls, look in their eyes). And I then asked her: "Do you know what a soul is?" And she had said: "Yes." "Have I a soul?" "Yes!" "Has a stone one?" "No!" "And a horse?" "Yes!" "A bird?" "Yes!" "And water?" "No!" "Have all dogs?" "Yes!" Lola had rapped this all out very nicely, and I praised her, to which she made response by a little spontaneous rapping—"isan..." "What does that mean?" "ich o wi glücklich!" (= I, oh—how happy!) "Because I am pleased?" "Yes! yes!"

4 December: To-day I said to Lola: "Why don't I understand dog-language?"[23] "Oft eil" (= often hurried.) "Yes, but even when I have tried, and paid attention I cannot understand!" "In hauch— z s u v z a e s " (= the first two words are "in breath," the remainder quite vague!) In a quarter of an hour I showed her a card on which a small child and a dog were looking at each other, and beneath—in Latin characters was written: "Wer bist du?"[24] "Can you read that?" I asked. "Yes!" So I put the card aside and said: "What is the second word?" "Bist." "But do you understand the sentence?" "Yes." "Which is saying it—the dog or the child? Look at both of them, they are young, and have met for the first time in their lives." "Both!"

11 December: "Lola! why do you and Frechi always bite one another when you are allowed to go loose?" "Ambitious!" "Ambitious to see who is the stronger?" "Yes!" "And which of you two is the strongest?" "Frechi!" She had applied the word with a nice sense of fitness: when two dogs meet for the first time this is exactly the

feeling that arises—either *one* of them is by far the strongest—a fact that both of them will be aware of, and silently acknowledge—or, their strength may be pretty evenly matched—in which case a fight will ensue, possibly even several fights, before the issue is finally decided. Is this not often *spiritually* the case between man and man?

13 December: Lola had been chasing after the game and had been punished by having to go without her food. She was however, in high spirits and rapped "esen!" following this hint in half an hour with "zu esen!" (= (give me) to eat!) I explained to her that this could not be done: that a punishment was imperative, if she would not break herself of her evil habits. Then Lola rapped out suddenly. "Lere mich artig sein!" (= teach me to be good!")

22 December: I have been showing her a picture in a book of Fairy Tales. My brother was present at the time, and it was the picture of the house of a robber, the house being drawn so as to represent a face: it had indeed been very cleverly executed.

"Lola," said I, "whatever is there about that house—do you notice anything?" (And thought she would rap "face.") She rapped. "Is a person!" I avoided looking at it again and merely asked, "Tell me, does it look friendly, or angry, or nice?" "Spetisch." "Spöttische?" (= mocking.) "Yes." And we both thought this reply admirable, for the "house" *does* look at one most "mockingly" out of the corners of its eyes.

31 December: "Lola, have you got worms?" "Yes!" "How did you get them?" "Ja, zige!" "An animal?" "Yes." "Is there a goat (= ziege) near here?" "Yes!" I had seen none about, but asked her again: "Where is the goat?" "Droif." "Do you know the name?" "Mittel!" (= her expression for anything she is uncertain about.) "Why did you say *droif*?" "I not any sort of word will give!" On making further inquiries I found that there *was* a goat in the immediate neighbourhood, and that the name of the family who owned it was *Freund*. I had never mentioned this name to Lola, so that she could only have heard it in the course of conversation among the people about, and then not very distinctly. In the evening, while I was absent, Lola stole some Marzipan. I expostulated with her in a serious, though friendly manner, and this evidently made her feel exceedingly un-

comfortable, for she suddenly rapped—"Sag irgend böse!" (= say something angry!)

1 January, 1917: "What is to-day?" "1.1. 1917!" "On this day we give good wishes to every one, so I will wish you much to eat, good health, and much going out: now wish me something!" "Am geln ..." (most indistinctly) I told her to repeat it, and she began again—"Am gu ... elen zu aufhören!" (i.e. am quälen zu aufhören = to cease teasing.) "You can't put a w after a g," I told her, but she persisted, and I waited in patience. There is no "q" in her alphabet, so she had found a way out very neatly! "Do I tease Lola?", I asked. "mich!" (= me!) This is indeed sad! and I am not conscious of my failing, indeed, I think that Lola has a very good time on the whole!

7 January, 1917.: "Now tell me something you would like to have explained, but mind you rap loudly and distinctly." "Ich o si so wenig kene." "Who is si?" "Dich!" (= thou!) (The reply had been "I know (or understand) you so little.") "Tell me what it is you don't understand about me? tell me something every day: what is it now?" "Work when I say no!" I tried to explain to her that my anxiety to get her work so lay in my desire for more knowledge about dogs—so that I might be able to tell everybody all about them, and thus make them kinder to animals. I took much time and trouble over my explanation, and at length Lola gave a responsive "Yes."

10 January: To-day we returned to the foregoing conversation: "Tell me what you don't understand about me?" "The food has also been worse lately!" she remarked. On this vexed subject I also attempted elucidation. I sought to explain the conditions of war, and that the amount of food available became less in consequence: that we people were no better off in this respect, and so on! And at length she again said "Yes!" Then I thought I would change the subject and asked her: "Why did Geri sigh so outside the door last night, and why does he look so unhappy to-day?" "er auch hat esen wolen!" (= he also wanted to eat!)

In the evening I said: "Lola, what *is* it you don't understand about me?" "Cause is often roughness!" She remarked—and here I really felt that there was little that I must needs explain—for I am not conscious of meriting her reproach on this score.

11 January: "Tell me something, Lola!" I pleaded. "Mistake to go out so little," she observed. Here she was emphatically in the right! She had not been out much lately, for it had been very wet—and she needs plenty of exercise. In the evening I invited her to "say something more." "o we gwelen!" "What worries you?" "ere nehemen!" (= taking honour!) "Taking honour about what?" "eid!" (So the old story has not yet faded from her memory).

12 January: "Well, now you've told me ever so much that you can't understand about me! But is there anything more?" "Zeig audawer (Ausdauer) in libe zu mir!" "*Ausdawer?* Isn't there a letter wrong?" "Yes, 4"; "What should it be?" "Au!" So the sentence ran, "Zeig Ausdauer in Liebe zu mir!" (= show constancy in your love for me!) Yes, indeed I will, you dear beast!

ULSE'S FIRST INSTRUCTION

As I have stated, when Lola came to me she could already say "yes" and "no"; she had even some slight acquaintance with the numbers and counting. The bridge leading from man to animal had been started, and the first difficulties embarked on. The further I pursued these studies with Lola, the keener became my curiosity to know whether I should be equal to the task of tackling this work where an animal in its primeval state was concerned, thus driving in the first props of this bridge myself! I tried my 'prentice hand in this work on Geri, the beautiful German sheep-dog, who had come into my possession in 1914. This dog—owing to excess of breeding, and also, perhaps, to the impressions imbibed in his youth was unusually shy and melancholy—he lacked all natural energy to "cut a figure" in any way; he had learnt to say "yes" and "no," and I feel sure that he understood me very well, but his nervousness and his constant fear held him back from rapping out anything beyond his *yes* and *no* answers. (At a later date I was obliged to give him away, owing to the scarcity of food.) Lola's progeny, therefore, seemed to offer more promising material for fresh ventures, but all—excepting the little lady-dog—Ulse—had been dispersed, going to their several new owners, before the winter days immediately after Christmas brought me sufficient leisure for further study, and as I had to give part of this time to Lola, as well as to the writing of this book, I had but a small margin left to expend on the little newcomer. Nor can I say, to tell the truth, that my interest in her was very great; she had already been promised to someone, and the fact of her still being with me was due to the difficulties of travel in these abnormal times. But, finally, sheer pity for the small creature—sitting alone in the stable—led me to bring her in for a few hours at a time so as to play about with me and Lola. One day it so happened that I had sent Lola off, and, being alone with Ulse, (mostly accustomed to intercourse with the maids) I attempted to teach her to understand: "Sit down!" To do this I pressed the little creature down on her haunches, saying, "*Sit down!*" And after I had repeated this three times she understood quite well what I meant, sitting down obediently at my slightest touch, and looking at me inquiringly out of her little bright eyes. I repeated this again the next day, and also touched her paw, saying: "*paw!*" Then I took the small paw in my hand and said: "Give a paw!" and in a few days this, too, had been learnt. I next taught her which was her right paw—and she very soon knew the difference. Indeed, Ulse seemed to think it all great fun, and was hugely delighted at the little rewards she earned. My interest, too, had now been aroused, and I repeated the numerals to her from 1 up to 5, and got her to understand "look here!" and "attention!" Though she was on the whole more fidgety than Lola had been, yet would she sometimes sit quite still, intent on watching my hand, but the least movement in the room would start her little head off twisting to and fro to every side. One day I took her paw, saying: "Now you must learn to rap! And placing the little pad on the palm of my hand, I first counted two with it, and then continued up to 5; then I held my hand out to her and said: "Ulse, rap 2!" and she actually did! I was delighted. I should add that before Ulse had learnt to "give a paw," she had already, of herself, shown inclinations to "rap," for she would hold up her paw—gesticulating with it in the air! These vague "pawings," moreover, were distinctly the movements of *rapping*, although she, of course, did not know their meaning at the time. And so the ground was laid for further work, during the short time I had to spare for her—as well as the limited period she was yet to remain with me.

There can be no doubt but that heredity plays a great rôle in these cases; her quick responsiveness bore witness to this, while, in addition, Lola evidently regarded her as the "flower of her flock," for she had always singled Ulse out for special attentions, generally retiring with her alone to a distant part of the barn. The question is whether Lola may not have given her some instruction, for, to some remark of mine, she had once replied: "Teaching Ulse!" Yet, for my part, I feel doubtful whether animals do transmit to others of their kind the things taught them by human beings. However, this may be, Ulse seemed predestined, so to speak, to learn to count and spell, mastering the numbers up to *five* in a fabulously short time. Moreover, she *rapped* better than Lola, or, rather, quite as well as Lola had done when in her very best days, raising her small paw high, and then bringing it down on my hand with a decided, though rather slow, beat. Ulse was also soon able to signify "yes" by two raps, and "no" by three, but I had to keep my questions within a very narrow limit, for her intercourse was of too short a duration to enable her to acquire a lengthy or varied vocabulary. Still, we practised 2 × 1, 2 × 2, 3 × 2, and her answers were always excellent, as long as nothing else was going on to excite or distract her.

The amusing thing was that she loved doing it so that the little paw would be up in mid-air as soon as ever she saw me, as much as to show that she was quite ready for work. This was doubtless due to the very quiet existence she had led before coming indoors, and also perhaps to the little favours and tit-bits she had learnt to associate with her new accomplishments. Indeed, until these had blossomed out, her innate cleverness and brightness had gone almost unnoticed.

When I had assured myself that she fully comprehended the rapping, I endeavoured to teach her to rap on a board, instead of on my hand, a thing I had never been able to get Lola to agree to. Indeed, I had had to relinquish any hope of it, in the case of the older dog; whether it was that the scratching of her toe-nails on the board irritated her or what, I do not know, but it practically stopped her working. My only reason for trying to introduce this method at all had been to put an end to the suggestions sometimes put forward by sceptical persons that I might be "helping her with my hand!" Anyway, the ease with which Ulse took to rapping on the board, and the excellent work she did by that method should have proved a sufficient reply to all doubters, and I had been full of hope that her gifts would, in time, have been further developed by her new mistress, yet it was to be otherwise. Ulse was to have gone to her new home in Meran (in the Tyrol), but the regulations as to travel obtaining during war-time prohibited this, so I placed her under the temporary charge of a young lady, and while there she unfortunately died of mange.

LAST WORDS

Everything that I have so far experienced or even heard of concerning dogs, I have attempted to set down here, and to do so has taken some fourteen months of close work. I have further added certain observations dating from an earlier period. It is my full intention to continue this work of experimentation, and should be glad if I might hope that what I have communicated in these pages may raise a desire on the part of some of my readers to embark on similar work in reference to other animals; for, in so difficult a field of discovery it can only be after much independent spadework has been done that the "complete form" we are groping after will be laid bare. Up to the present it may be thought that little of really practical value has been proved, and to some this may suggest that the work is therefore superfluous. But, do we study astronomy for mere *practical* reasons? Does the seeker in this field of science imagine that he is going to derive *practical* results for us, *in the immediate future*, from his study of the heavens? It is for purely *ideal* reasons — and in order to give seeking humanity that which is indeed theirs, that we humans send forth our thoughts, exploring every region of the world — be this "of use" or not! And in thus probing the depths of our own subject do we not come up against those weightier questions which are of Cosmic importance? Does not Nature here fix man's eye with her own gaze — granting him new riches? For rich, indeed, is this gift that proves to him that not he alone is dowered with a soul[25] — nor dwelling in a world destitute of thought, nay — that his companion-beings along life's highway are well able to respond to and comprehend all his labour, his love, and his care for them. And above all, should it teach him to more clearly apprehend them — doing so in the spirit of a know-er and with a kindly sympathy begotten of that knowledge. For *To Know — to Understand —* means to give to each its rights! And, in this matter, have we to concede so much to our higher animals? The simplest form of thought contents them; the childlike adapting itself to animal uses; and, from such "small beginnings" has not our own primeval soul — the best that is within us — risen to higher glory, to become a moulder and organizer of thought — even of creative ideas? Therefore, from all that wealth with which we are dowered we may well allow this tiny morsel to our animal friends — they will assuredly infringe no further upon our rights, for, after all, they are dumb, and cannot even utter the small store of thoughts they may learn to express; they can only look at us—but, oh! how well they can do *that*—it needs no more than our eyes to tell us! And—if we review the entire animal kingdom, are not these *higher animals* closely akin to us, both in bodily structure as also in all that appertains to their functional activities? So near, indeed, do they approach us in the degree of evolution that for that very reason it would seem natural to attribute to them some rudiments of thought—some latent abilities; but the greatest importance of all would seem to lie in the Cosmic aspect of this question! If it *does* "fit in" ought we, then, to dismiss it? Is it not the same thing with all subjects that open up a new point of view? Yet may those for whom such new investigations present no "disturbing elements"—those for whom, on the contrary, it chimes with their own desire—extend their hand and gratefully accept this gift from Nature—repaying her with reverence and with love. May this new science serve to enrich our ever increasing knowledge! The work will indeed mean a long struggle against the conservative elements, and all those accepted rules of procedure; every weapon will be turned against us, but, be this as it may, time will in its due course show the truth to be on our side, for ONLY WHAT IS TRUE SURVIVES.

CONCLUSION

By Professor H. F. Ziegler

The most important contribution that had been made to the study of Animal Psychology consists in the new "Alphabet of Raps," which enables dumb creatures to give reasonable expression to their thoughts, and provides us at the same time with the means of gaining some insight into their thinking and feeling. This method owes nothing to scientific investigators, yet may these gladly acknowledge the great progress thus indicated, rather than reject it with impatience and distrust. To proudly decline anything to do with it would indeed be out of place: rather is it careful study and independent confirmation—a personal application of this new method—that is here most needed. The inventor of this "Rapping and Spelling Method" was the late Wilhelm von Osten, in Berlin, reference to whom has been made in the opening chapter of this book. But the specialists refused to recognize his labours—they destroyed his position by their erroneous findings and their disapprobation—the campaign carried on against von Osten being by no means free from a spirit of unfairness.[26]

It was Karl Krall who took up and continued the work, improving on the original method and finally making known the most astounding results which he himself had succeeded in obtaining with his horses. These accounts may be read in detail in Krall's great book, a work the publication of which has been of immeasurable importance in the history of animal psychology.[27] Any reader of unbiased opinion will be bound to acknowledge the value of this new method, and the remarkable results achieved in the case of Krall's horses have been equally successfully applied when working with dogs. Frau Dr. Moekel of Mannheim evolved an independent rapping method of her own, which admitted of the possibilities for *counting*. This lady, however, soon became aware that a similar method had already been invented and applied by Herr von Osten, and she then enlarged on her own efforts so as to include the spelling method above mentioned. The feats of her dog Rolf were so remarkable as to arouse as much surprise in his mistress as in anyone else present. Frau Dr. Moekel was exceedingly careful to note down everything that could serve as evidence, and in spite of her long and serious illness was yet able, by dint of great exertion, to complete her MS. She died in 1915, and her book, which could not be published during the war, has only recently become available to the public. It is gratifying to be able to welcome the appearance of another little book on the same subject, the one now before us, written by Fräulein Henny Kindermann; this volume having also suffered postponement, owing to the war. This lady taught her dog on independent methods of her own, devoting much loving and conscientious care to the work and, in a general way, the results have been much the same as those obtained from Rolf, although, in the matter of detail, there is much that is new; indeed, many of the observations set down by this investigator raise questions of fascinating interest. Here, again, the author has been able to improve on the method as previously applied by others; teaching the dog to rap tens and units with different paws, as had been done by Krall's horses, and also introducing a better method of spelling by teaching the proper value of the consonants.[28] Fräulein Kindermann further applied her tests systematically in order to solve certain problems, proving the animal's ability to the full extent in one particular subject at a time. It is indeed the experience thus gained which gives to this book its special value, even though all the problems submitted may not have been fully solved. I would here draw attention to the fact that the author's dog invariably replies in "High German," whereas Rolf of Mannheim employs the dialect of the Pfalz—and the Stuttgart dog, Sepp, expresses his views in Suabian; indeed, each dog naturally learns the "form of speech" he hears in his own locality. The results that have come under notice seem at times so extraordinary that doubts may arise as to the authenticity of what has here been set down; yet should we be careful not to reject new evidence because it happens to exceed all we have hitherto known or experienced. For this is a case of exploring new ground, ingress to which has now become possible owing to an entirely new method, and none should take upon themselves to decide in advance what may, or may not be, found possible within this new domain. Careful examination of all evidence put forward is desirable, yet can this be undertaken only by such persons as are themselves in the possession of an intelligent dog, one to which they can apply the test of similar instruction. It should be needless to say that the experimenter must abstain from anything in the nature of a sign given to the animal. It is a far easier matter to train an animal in *that* way than to bring out the latent possibilities attaching to its understanding by training it so as to state its own thoughts. The proof of the genuineness of such "utterances" on the part of the dog lies in the fact that it so often gives an entirely different reply to that which is expected of it—it may even say something that is quite unknown to the person carrying out the experiment. Many such examples will be found in this book, as well as in that of Frau Dr. Moekel, while many more could be furnished by the owners of other "Spelling Dogs." Indeed, the more reckoning and spelling dogs there are the sooner will the value of this new method become generally recognized and the easier will it be to rid the truth of any errors that may still obscure it. Here in Stuttgart my Lectures delivered on the subject have so far led to the training of four dogs in counting as well as spelling, this having been done with best results. In addition to these, I myself have a dog, "Ava," by name a daughter of Lola, who is already proficient in both accomplishments. There is nothing mysterious about this new animal psychology that has been brought into evidence by the method here explained, it is no secret, but at the service of all who care to explore what is entirely free ground—not reserved for the learned alone, but at the disposal of any animal-lover, if he will but co-operate in a spirit of patience and devotion, and is endowed with the particular "gift" for teaching an animal. The truth under discussion here is not likely to be find elucidation in the study of the learned man—rather will it be the result of the collective, convergent and corresponding evidence brought together by the labours of many a patient investigator.

Stuttgart
September, 1919

NOTE

There are in all now twelve dogs known to communicate by means of "raps." The experiences I have had with my own dog have been reported by me in the article entitled "Respecting a Dog's Memory," and appeared in the "Zoologischen Anzeiger," 1919, No. 11-13. The name of my dog "Awa" is quite intentionally put together, as Lola has herself "invented" all the names given to her progeny.

"THINKING" ANIMALS
A Critical Discussion of Developments from 1914 to 1919

BY
DR. WILLIAM MACKENZIE
OF GENOA

[Translated from the Italian with the omission of
I. An Introductory Section, and
II. A Section giving the Story of "Lola."]

III. THE HYPOTHESIS OF INTELLIGENCE IN ANIMALS

Assuming, as I have done, and as I think I must do, that we have not here to do with a trick or fraud, we seem to be dreaming, or to be reading the account of a dream. Those poor horses of Elberfeld, so greatly extolled and so much discussed in their day, are not in the same field with Lola. And yet I am convinced that it is not a dream. It is another kind of psychological reality, but it is a reality probably too complex to be reduced to a single formula. Let us then try to face the facts.

As to the "intelligent" character of the manifestations, there is no possible doubt, even though we put on one side for the present the arithmetical phenomena, which perhaps must be treated from a particular standpoint, as I shall explain. The question before us is therefore a dilemma. Is there intelligence in the dog, or is the intelligence in others?

If, by intelligence in this case we mean the possibility of the animal under observation giving replies to questions with, in the human sense, actual understanding of the import of such replies, as well as the possibility of the animal, a dog two years old, being able after a maximum of fifteen hours' lessons to read, write and count, *and know what it is learning*; if that is what is meant by intelligence in this case, I must say that I do not believe in it, and that I feel compelled for scientific reasons to examine every other hypothesis before having recourse to this one.

And again, "Intelligence in others"? This may be so, but it is not necessary to suppose that the intelligence is in others alone. I mean that a few of the manifestations may within narrow limits probably be rightly attributed to the intelligence of the animal, (but, I repeat, the arithmetical facts must be considered by themselves).

If all the manifestations were to be attributed to the intelligence of others and none to the animal, we should have to accept the supposition of an absolutely *mechanical* automatism in the animal itself of the type suggested by Neumann (8)[29] as the result of his experiments with Rolf, when, for instance, the dog mechanically kept on tapping an unlimited number of times on the cardboard, which Neumann held out to it without, as far as possible, moving it.

This negative result of Neumann's is capable of various possible explanations, and in no way gives any clear indication (just because it is negative) as to how a positive result is at all possible; that is, we cannot conclude from it any better than before, whether the apparently "mechanical" behaviour of the animal was intentional, and therefore whether the animal itself could or could not have behaved otherwise; whether, given the impossibility of the animal behaving differently, we should say that this impossibility was absolute or only happened to occur on this occasion; whether perchance the action of some psychical factor unknown to Neumann between the animal and himself may not have been omitted; and whether such factor was not in operation when the animal was working with its late mistress, etc., etc. In this connexion I feel it incumbent upon me to recall that I myself saw Rolf on two or three occasions behave in this same apparently mechanical way with his mistress (Mrs. Moekel) (II), whose annoyance thereat seemed so real that I felt certain that it was not feigned. From Neumann's point of view this would be incomprehensible—since he makes use of the argument from the supposed absolute automatism under the impression that it had taken place in Rolf with *him*, Neumann, alone, *but not* with the Moekels. Here, then, it is clear that the intelligence is, or at least that it is also, "in others."

But whatever value we may attach to Neumann's experiment, it appears to me sufficiently clear that the supposition of an absolutely mechanically passive process in the animal will not hold as a sufficient explanation of the *whole* of the facts related by Miss Kindermann, nor will it hold with regard to what science certainly seems to me to be compelled to admit in the case of the Elberfeld horses, which (as is known) "worked" magnificently without contact with anyone, tapping their replies on a board, completely isolated on the ground, and even when all alone in their stable with the one door tightly closed and all the spectators outside. The spectators heard and observed the rapped answers of the horses (for example, to written questions) through a little glass window. Neither will it hold with regard to the many experiments made, some also by myself, by means of requests, pictures, questions, presented to the horses in such a way as to be unknown to *everyone*, including the experimenter. Besides, the animals at times gave spontaneous communications. This Assagioli and I, and many others, have observed even without the presence of Krall and of members of the Moekel family. Miss Kindermann also gives some of Lola's replies tapped on the arm of a friend of the authoress, although the latter held out as usual her own hand to the dog.

Therefore, there must be some "intelligence" in the animal, as everything cannot come from outside it in these experiments. Probably this intelligence is not human in quality, but nevertheless not quite rudimentary, and is such as we may imagine without too much effort to exist in domestic animals which by many signs often give us proof that they understand at least in part what is taking place around and within us. That such an intelligence could very probably be educated, always within prehuman limits or in a lesser degree than in human infancy, does not on the whole seem to me so contradictory to our actual psychological knowledge: since we may very well suppose that the animal under examination may make use of its proper faculties, as far as lies in its power, to profit by the situation for the purpose of accomplishing that which is required of it, under the stimulus of allurements or threats. (It may even be rather assumed that the exercise of its proper faculties, which I regard as "intelligent," may procure for the animal a certain degree of pleasure.) All this is apart from the question of the arithmetical phenomena which, as I have already said, deserve separate consideration.

Upon the facts as now established the knowledge of numbers seems to be the basis of any educability in animals. And this is perhaps the first and most important discovery in the "new zoopsychology."

In their search for others things, Von Osten, Krall, and the Moekels have brought out clearly among various other facts, without exactly accounting for it, the fundamental fact of the existence in the animal of a psychic substratum predisposed in some manner to arithmetic. I say "in some manner," and by that I do not wish to prejudge any particular view of the argument; and above all I do not make of this predisposition or mathematical permeability, a criterion of intelligence. I do not forget either the mentally deficient or the prodigies among child calculators, etc. But likewise I cannot forget another thing: that all organisms are already throughout permeated with mathematics, and that the more we descend the scale, from man down to the most "simple" biological fact, the more nearly we approach to physics, which is nothing but mathematics.

I have not the space here to digress on the intermediate gradations. Besides, I have already done so, in part at least, elsewhere. But I wish to recall the curious coincidence that the mathematical achievements of the Elberfeld horses were much more brilliant and much more prodigious than those of the dogs which have up to now been experimented on. And horses in the phylo-genetic line are more ancient than dogs: they are lower in the zoologic scale. Much lower still, i.e. among the Arthropoda, occur many other mathematical wonders. I only mention in a cursory way the logarithmic spiral of the spider's web, the precise curves realized without instruments of any kind by the Coleoptera and Hymenoptera in cutting leaves, the stereometry of the aphides. Then, as it were, at the bottom of the scale (if one may still speak of a descent and a bottom) the marvellous plancton filters of the Appendiculata; the geometrical spots of the Amœbae; the cases of perfect forms of so many other Protozoa; and, finally, think of the constructive technic of the static organs, or of those of movement either in man or animals or plants; think of the complex mathematics of the mitosi, or of any cell proceeding to its own indirect division.

It seems to me clear that the mathematical faculty — assuming always, let it be understood, that it may give rise to more or less conscious phenomena in the biological subject — may be amongst the most natural of imaginable causes, and that even the smallest amount of consciousness may help this existing capacity in the animal to express itself. That we are concerned with an expression by raps or not, does not seem to me as important as a proper estimation of the importance of the central fact constituted by this mathematical capacity.

From this central fact, proved over and over again without any possible doubt to be true of the "thinking" animals, there have been developed two distinct groups of consequences: (1) the prodigious mathematical performances occurring as by magic among the Elberfeld horses at a certain point of their "education": (2) the apparent manifestations of thought through the typtology or rapping out of words, culminating in the "philosophic" achievements of Rolf and Lola.

For the reasons just mentioned the first group of consequences seems to me to admit largely of biological (i.e. biopsychical) explanation; however, anything which eventually does not fit into the biological explanation may be made to enter without any effort into the second method of explanation which, in view of the facts, it seems to me that we must adopt for the second of the two groups of consequences above referred to.

That mathematics can be "lived" rather than "known" — or, if any one prefers the term, "realized" — by an organism which is without any psychical accompaniment whatever of the human type, is a fact which I find credible. But when Rolf speaks to me of the origin of the soul, or makes up poetry; when Lola complains to me of honour lost, etc., the thing is not credible to me in any way except by paying attention to nothing except the feeling, which is so difficult to avoid, that what is here speaking to me, versifying and complaining, is a psychical "quid," absolutely human and only human; a "quid" which therefore is (after all) not the animal's, although manifested in some way through it. The difficulty naturally consists in deciding precisely how this happens. But it does not seem to me altogether impossible to arrive at a proper hypothesis.

I have already said that we must discard, because of its inability to explain a great part of the facts, the most easy and simple hypothesis — that of some mechanical signal (e.g. by means of a supposed pressure of the hand under the cardboard, or by the hand itself which is held out to the animal, in the case of the dogs which have so far been experimented with). Here we also have to remember the proposition laid down by Miss Kindermann herself that "She did not wish to let herself be carried away by sentiment," and that she would seek all possible proofs which were good logically. Having excluded the hypothesis of deceit, it is a further proof of the sheer impotency of the theory of signals, when regard is had to the available amount of the material observed and recorded in the authoress, if we ask how is it possible to imagine that she (knowing very well, as she says, the suspicion resting on the method) in a year or more of work with Lola should not herself have perceived that she herself had been producing by mechanical means the rapped answers of her pupil?

In my opinion the answer is that the authoress was not only not aware of, but *could not* in the least have been aware of, the action that may have passed from herself to the dog so as to bring about the rapping of the answers; and that on the other hand it is not a question at all of thinking of a simple mechanical operation of the kind mentioned above, because in the presumed action of the authoress on the dog there is no need to have recourse to such a crude hypothesis (as surely there was no similar action of Krall's on his horses, especially when they were separated from him). I maintain, in fact, that in principle, even without any contact by hand, we may still presume that all the "wonders" obtained by Miss Kindermann are obtainable, taking, of course, into account the peculiar endowments of the animal we are dealing with. For if there be any automatism (and there is surely a good dose of it), it is certainly not a question of a mechanical automatism (of the type of Neumann's), but quite certainly of a true and proper *psychic automatism*; a very different thing, and without doubt much more complex.

In all probability the first condition for the occurrence of genuine phenomena similar to those attributed to "thinking" animals must be a very particular psychic relationship between the animal and his master. And such a relation, although with reluctance, I am compelled to call of the mediumistic type.

My reluctance is due in part to the very unhappy etymology of the term, derived from the famous word "medium," so unscientific both in its origin and in the meaning which some even now wish to associate with it. But even after having freed it from any "spiritistic" meaning, the term still leaves me reluctant; for I cannot hide from myself the weakness of a hypothesis which, in order to explain (only in part) one enigmatical fact (in this case, that of "thinking animals"), must have recourse to another unsolved enigma (in this case that of the "mediumistic phenomena").

However, it will already be something if the two problems are eventually merged together and so become a single problem; but it is not my object to explain any psychical facts themselves, whatever they may be, under which the phenomena of Lola and others of a similar nature may be eventually classified. It will be sufficient for me at present to group the performances of the animals, if possible, with something better known. And "mediumistic" facts, extrinsically at least, are certainly better known. I refer therefore to them as I find them described in the psychology called supernormal; because, from force of circumstances I am compelled to recognize that it is within this psychology that I must now continue the discussion.

IV. MEDIUMISTIC "RAPPORT" AND TELEPATHY

The hypothesis of a psychic automatism of a mediumistic type, as a concomitant phenomenon, at least, in experiments of the "new zoopsychology," offers us a point of support for a possible interpretation of the strange uncertainty and irregularity of the successes and failures of different observers and different animals.

With Krall two of his horses gave magnificent results; two others negative results. In the same way, with the same dogs some experimenters obtain wonders, others obtain nothing.... We may therefore assume that in order to obtain favourable results there must be a proper accord or reciprocal psychic concordance between the animal and the person making the experiment, precisely as happens with mediumistic phenomena.

Moreover, this hypothesis in the same way helps us to an interpretation of the fact that the same animal, with the same investigator, gives good results in some matters, poor or no result in others. Taking, however, due account of the central mathematical phenomena, on which, as it seems to me, the whole edifice is superposed, there remains a great variety of marked psychical idiosyncrasies in the various cases. One of the animals is decidedly a calculator; another likes to read or to explain figures; another detests reading but willingly taps out "spontaneous communications."

Without possessing much intrinsic probative value of its own, it is certain that all this fits in very badly with the supposition of a purely mechanical automatism operated by the person making the experiments. And on the other hand it bears a close analogy to the mediumistic "specialities"; that is, to the well-known fact that one "medium," for instance, is good for "physical effects" (i.e. gives rise around it to dynamic phenomena), but is not good for "psychography"; or produces "incarnations" but not "apports," etc. In the same way, typtology or rapping, more or less systematic, seems a fundamental gift, common to all the various kinds of "mediums." And the fact is perhaps of a certain value that precisely the same thing is true of "thinking" animals; although we must always remember that an analogous relation may only be apparent or extrinsic. Besides, the tone also of the "communications" in the two fields seems to me very much akin. I allude to the curious, angular, enigmatic, spasmodic, often playful and bantering communications, with frequent "unexpected replies" and philosophic platitudes. I find all these in Lola, and I remember similar stories of Rolf and of the horses, giving me an impression very like that which I get from the accounts of mediumistic seances "with intellectual effects."

Premising all this, we may suppose that a peculiar psychic concordance, which failing a better term might be called mediumistic, exists between Lola and her mistress. The mistress then in some way will have "communicated" through the dog the substance of her psychic self (perhaps with eventual autonomous additions from the canine or other psychic entity); all this happening, we must suppose, in a subliminal way, with partial psychical disassociation on the part of the authoress, if not also probably on the part of Lola, about which I am quite certain (and in this I agree with Neumann) that it absolutely does not understand anything or know anything of almost all the manifestations of thought which it exhibits.

There remain the questions (if the possibility of such duplicate mediumistic phenomena is admitted *a priori* to be possible) as to the point at which the normal relationship between a human person and an animal passes over into this supernormal one; and, finally, as to what particular known facts in the case of Lola, besides the rather too general analogies already mentioned, speak in favour of this hypothesis.

Into the mediumistic endowment of the investigator it seems to me useless to inquire since *a priori* many persons, so it seems, are more or less strikingly endowed, and the conditions which determine results are not sufficiently known. At the most there exist some indications — e.g. in Morselli's masterly work (2) — of the existence of some concordances between the phenomenology of mediumism and hysterical, hysteroid, or at least "sensitive" temperaments. And I believe that — with the help of their own publications, properly analysed — it would not be too difficult to attribute one or the other of such physio-psychic varieties to those persons who have up to the present obtained the best results with "thinking animals."

More interesting appears to me the investigation of the question whether animals themselves have already given any clear proof of being able to be "sensitive" in the mediumistic sense. And I must say that such a proof seems to have almost been reached.

I may refer on this subject to the exhaustive monograph published in 1905 by Bozzano (1) and written with the special competency and clearness that distinguish the well-known Genoese psychist.

Bozzano at that time was necessarily ignorant of the "thinking" animals, for it was only afterwards that they came to notice. But there were other authors who introduced the possibility (or the necessity) of a supernormal relationship in order to explain the Elberfeld facts, as soon as they were known. Perhaps the first in chronological order was De Vesme, who published in 1912 an interesting article in that sense (3), showing the many analogies between the phenomena of Elberfeld and mediumistic phenomena generally, e.g. the typtological particularities; the wrong orthography ("Firaz" tapped by the horse to express its own name "Zariff," "Dref" instead of "Ferd," etc.); solutions of difficult problems and invincible resistance to simple inquiries; immediate promptitude of correct replies to complicated mathematical problems, etc.

A similar work was Maeterlinck's, written in 1909 for a German review, and then transformed into a long and interesting chapter of the well-known volume, "L'hote Inconnu" (10).

Then in 1914 was published a book by E. G. Sanford (5) containing some useful comparisons between "thinking" animals and mediumistic psychology.

In Italy there were indications in the same sense, in the work of Stefani (1913), Professor Siciliani (1914), and others. But the subject was but little followed up.

Even psychologists by profession seemed for a time to be willing to accept the hypothesis of some "telepathic" transmission of thought from the investigators to the Elberfeld horses.

Already Claparede (1912) had been forced to refer to this, although he refused, so to speak, to discuss the matter; then G. C. Ferrari, and F. Pulle, in an interesting account (4) relate how the horse taken by them for instruction sometimes guessed the numbers that they were proposing to them, and rapped out the answers before being asked to do so.

Whatever may be the fate of the telepathic hypothesis, it may not be amiss to remind the reader that it undoubtedly is very closely connected with the mediumistic. The distinction between them is not always easy; besides, both may exist together side by side.

"Telepathy," so called, (a term not less unfortunate than that of "medium" and its derivatives), or, better, the transmission of thought, is (shortly put) the hypothesis that at a certain moment an agent transmits, and a receiver perceives, some definite mental image or state of mind. The transmission may be more or less willed (i.e. conscious) on the part of the agent; on the part of the receiver, however, the fact of the transmission always remains unconscious, but the psychical elements perceived bring about a reaction in consciousness and the receiver knows what he is doing, or at any rate may do so, at the moment of the occurrence. Shortly stated, it may be regarded as a kind of suggestion, "à distance," with sometimes immediate and sometimes delayed effect; a kind of posthypnotic performances of a suggestion without the intervention of hypnotism (or, perhaps, with a partial subhypnotic state?), the receiver of the suggestion not receiving it in the form of acoustic vibrations or in any way by means of one of the ordinary senses.

Mediumistic phenomena on the other hand require for their explanation the possibility of a much more direct, more profound and more immediate relationship between the several minds taking part in them. One of these minds—more or less disassociated—might become the instrument of another—even of several others—although still itself in a state of more or less complete disassociation, and always remaining altogether unconscious of its relationship to the other. One of the minds might therefore be an agent, another a recipient, or even several of them simultaneously might join together to produce the phenomena, the subliminal nature of the relationship remaining fixed. The actors would in this way, for ever, all of them without exception, be absolutely unaware that they were the actors. It might also be the case that the recipient through whom the phenomena are produced (i.e. the "medium," or in our case the animal experimented on) would not be conscious at all of the resulting action. With human "mediums" we should find in such cases a more or less advanced state of trance or ecstasy. And with regard to animals, I remember the opinions of Ochorowicz and others—which were preceded, however, long ago by a similar opinion of Cuvier—according to which the consciousness of animals in an awakened state would correspond fairly closely to the consciousness of man in a hypnotic state.

If what has been said above is at all correct, it would seem as if the walls separating various minds one from another all of a sudden are opened wide, and by a partial interpenetration of one mind by the other the several minds join together to produce by mutual determination automatic action. And it is in these special psychical states that "supernormal" phenomena, viz., psychography, clairvoyance, clairaudience, etc., occur.

Now, although all this is to move in a very uncertain ambit, harassed by a multitude of diverse and vain dilettantisms and mysticisms, and only too frequently by fraud, it is not any longer possible nowadays to deny that facts, objectively known, compel the positive scientist to have recourse to some such suppositions. Also without making the "subliminal," with Myers, a kind of "deus ex machina" in the world, it is certain that mediumistic phenomena of the kind mentioned are henceforth to be considered as a subject of study for an open-minded psychology. I may refer in support of this view, among others, to the powerful work of Morselli. And to return to the "thinking" animals, we find that the mediumistic hypothesis, however shifty it may seem, is a better explanation than the telepathic hypothesis—which has already itself become rather more systematized in modern psychology.

After his visits to Elberfeld, Claparede, as I said, had found it difficult to treat as valid the telepathic hypothesis when applied to Krall's horses. What, indeed, had been "transmitted" to them? Numbers? Words? Single letters? (or orders to stop the foot at the right time?) It must be remembered that the horses were tapping their answers by using a sort of stenography, that usually left out the vowels: that besides, although the words could be recognized in the most certain manner, the spelling was most irregular, and, as I have already pointed out, sometimes reversed. Further, as to the words themselves, most infantile phrases were used, certainly such as no adult would have suggested. Was it suggestion then from one unconscious to another? But this is to fall back upon a supposition of the "mediumistic" type, and takes no count of the cases of replies to questions which were unknown to everybody present, and brings us to the single dilemma: either there is intelligence in the human sense in the animal, or a relationship of the mediumistic type above described between the several minds concerned.

As to the interesting observations reported by Ferrari and Pulle, it seems to me opportune to quote here some extracts from the first of these distinguished authors.

"This séance was particularly interesting, because I find it recorded in my notes that a fact was verified three times consecutively, which had occurred sporadically more than once before, and had been observed and noted by us and various other witnesses.

"It consisted in this: While I was putting in the box the number of balls which I had intended the horse to read, the horse, which often could not even have seen the number of balls, because I covered them partly with my head and hands, tapped out the correct number.

"The same thing happened when I took in one hand a card, the signs on which it could only have read with difficulty, the light being rather bad. The most curious thing about it was that the taps were then made upon the whole more rapidly and less strongly than usual; and that several times later on the horse gave the same number itself with some little difficulty.

"It is also curious that it should have repeated the performance, seeing that it was only once rewarded for it, and that, because it was agreed that it had done its reading well. I must add that the person who assisted me told me that generally, even when it was giving correctly the number decided on, it hardly looked to see how I was placing the balls in the box....

"Once when I was arranging three balls, because some one standing behind the horse had made me the sign 3, the horse tapped its three beats behind my shoulders while stretching out its neck by my side in order to try to take a salad leaf, thus showing that it was taking very little interest in the sign which I held out to it and in the taps which it was making.

"Certainly, this time at least, the animal seemed to perform an automatic action, and it seemed to me that we had guessed subconsciously what the horse intended to do. This may appear a crooked hypothesis, but it is less difficult for me than to think that the horse had read in my mind the number which I had there. It certainly did nothing on most occasions to upset the fairly clear and precise impression that it was obeying some more or less complex determinism."

It seems to me difficult to avoid the impression that what has just been stated does not reveal a simple telepathic relationship but something rather more deep. The want of interest by the animal in its behaviour is for me symptomatic, and agrees perfectly well with the sensation of the observer that he also had to obey some obscure determinism. I see here another case of a combined psychical (partial) operation of a "mediumistic" kind; and this hypothesis makes very plausible the other no less impressive hypothesis of the observer that his mind was reading (in a subconscious way) the mind of the horse. I call this hypothesis of Ferrari impressive, because in this case it was due to a person who is certainly not to be suspected of dilettantism, and still less of any pseudo-scientific mysticism.

For the rest I repeat that "telepathy" also may co-exist along with "mediumistic" action. In a general way, telepathy would seem to assume in the animal a greater amount of "human" psychic affinity, whilst in mediumistic action I look upon the animal as reacting to the intervention of the other mind in a much more "automatic" way: almost like a "speaking table," but a table provided with live feet rather than inert legs, and above all provided with a nervous system forming part of it, so that very little action on the part of the medium is required, but the subliminal action of the investigator is enough by itself to work it. (Of course, this does not exclude altogether action by others or by the horse itself).

Krall admits the possibility of telepathy (but in a very limited measure): and then, if I remember right, he was looking finally for an explanation which to-day I should perhaps call of the mediumistic type, if I had been better acquainted with it; but in fact I had of him, in his lifetime, only some vague hint on the point.

As to Miss Kindermann, she recognises the possibility of transmission of thought in certain cases (e.g. when Lola is tired or is unwilling to "work" any more). According to her it would be a question of a line of least resistance, along which the "work" of the animal becomes more easy. Hence arises the necessity, as she maintains, for the investigator to be very careful of the danger of falsified results and to *abstain with this object from any intentional thought*. But these are the very conditions which "mediums" impose on investigators, and if these conditions are not observed, mediumistic séances seem only to be successful with difficulty. Therefore, in trying to resist the danger of telepathic falsification, and without indeed being aware of the resulting consequences, Lola's mistress may have contributed to create the very conditions most favourable to the development of mediumistic action.

V. THE HYPOTHESIS OF CONCOMITANT PSYCHICAL AUTOMATISM

In various parts of her book Miss Kindermann emphasizes the fact that after having given for some days "communications" of a certain kind, a sort of tiredness or annoyance, that gets hold of Lola, completely prevents the repetition of similar communications; but that repetition can take place if some weeks of rest are allowed in the subject which has provoked the tiredness.

In another place she mentions that, with the progress of Lola's "education," the dog's attitude towards herself, and other persons generally, became harder and more difficult, almost hostile (a fact which I find confirmed by certain answers of Lola's referred to elsewhere); just as if the canine consciousness as it gained illumination began to understand the many wrongs done to it by man, which formerly it knew nothing about.

Other observers have repeatedly stated that a capital fact in the story of "thinking" animals is the necessity, which they regard as proved, of a *progressive* "education" directed at getting from the animal results proportionate to the instruction received.

All these observations and several others of a similar nature would seem to be arguments in favour of a presumed "intelligence" rather than of an automatism in the animal. But they should be accepted *cum grano*. They may indeed contain a good dose of involuntary suggestion, active or passive. And again, it seems to me, for instance, a very doubtful procedure to maintain, after a positive result has been achieved by the animal, that the result should have been on the other hand negative, if the education has not yet reached the corresponding stage of development; and vice versa. As for me, when I read what Miss Kindermann writes about the rapidity of Lola's progress, I cannot help thinking that, if the authoress had believed that she was able to obtain at once from the dog the results which she did obtain after a year's work, she would have obtained them fully and completely.

But this extreme supposition may be exaggerated. I have already repeatedly referred to the hypothesis that the psychic automatism in question may be only concomitant. That is, I am convinced from what I have seen myself and read that a foundation of intelligence, of logical reasoning and of self consciousness, must go to constitute in the animal the substratum on which the wonders of the "new zoopsychology" are built up.

At first I was rather inclined to believe (as so many others) that the facts discovered at Elberfeld and at Mannheim could and should be explained simply by the recognition of "intelligence" in the animal. The chief results obtained up to then (i.e. up to the date of my last publications on the subject), were the mathematical prodigies performed by Krall's horses, and the first "philosophic" manifestations of Rolf. I accordingly thought that I should be able to interpret the new (and, in its complexity, rather modest) canine "knowledge" by the animal's memory of words which it had heard. But since then the educators have taken pleasure in raising the whole level of these wonders. Rolf's "philosophy" was developed; and in the end they went so far as to make him compose poetry, as I have already had occasion to mention. Then came the performances of Lola. And at this point I, too, must say: "Too much, too much!" At least, as far as concerns the hypothesis of intelligence in the animal.

I understand perfectly that just on account of that "too much," people may be tempted to throw up the whole thing. But as far as I am concerned, I repeat that I do not consider myself justified in doing so. I do not forget the possible intervention of active or passive suggestion: I referred to this a short time ago. But a great abuse is often made of this explanation. In practice "suggestion" explains but little to any one who wants to get to the bottom of things. Neither does it explain the bulk of the facts of the "new zoopsychology." Neither do I forget that in this field also (as in every field of psychological experiments) there may be an interfering although subconscious misuse of spurious factors, such as signs (not intentional or perceptible) by the experimenter to the subject experimented with; a certain amount of falsification in interpretation of results on the part of the experimenters, etc.... But the irreducible residue of the facts is, in my opinion, still enormous as compared with the little that could perhaps be eliminated by these means from the discussion. Therefore, in the absence of anything better for the moment, and subject to further information, I hold to the hypothesis of a psychic automatism of the mediumistic type, as a concomitant phenomenon developed from the normal "rapport" which is *necessary* and pre-existent.

This "rapport" is that of a master to a child; but to a very special kind of child, a "child" moreover who, from the biological point of view, has not been corrupted by the thousands of years of reasoning and society that weigh on the human child. It is, therefore, nearer to the "fountains of life" if I may be allowed to express myself in that way; and nearer to the mathematical potentiality (which was at first unself-conscious, but which has subsequently been developed). But, of course, it is not enough for mathematics "to be" in something, for that something to begin at once to tap numbers. The table of the mediumistic séances contains much mathematics (in its physical assemblage), but in order to make it "tap" there must be somebody to move it: in fact, a "medium." In my view, as soon as the animal subject has been able to understand "numbers"—and this postulate of the new zoopsychology, I repeat, I believe to be indispensable to the whole edifice—the animal finds itself sufficiently in harmony with the master to become capable (in principle) of all the subsequent "wonders."

This it is which constitutes the first discovery, as I have called it, of the "new zoopsychology." And on that discovery, in my opinion, are based through various gradations its chief results, on the supposition that at a certain moment there takes place a new specific action, the "déclanchement" of the mediumistic relationship between the animal and the experimenter. And it may be that the development of such a very special relationship between man and animals may be comparatively easy. That is, it may be that the animal is relatively easily *permeable* by a mind provided with a reasoning intelligence (without, however, being itself aware of the logical content of such an intelligence), exactly because it is rather poor in logical self-conscious content—or, again, it may be, that the animal in a certain sense is nearer than we to the "fountains of life." (9).

The possibility of this "déclanchement" would therefore constitute the second and more serious discovery made by the educators of animals; although without their knowing it, as is proved by all their accounts which make no mention of it.

It is difficult to say what the precise moment is at which the grafting of this supernormal connexion on the normal one takes place. The most that I can say at present is this: that the grafting in question appears relatively to be quicker as regards the mathematical results. And this would lend an indirect support to the view that generally mathematics must be presupposed as underlying the phenomena. But the wonderful performances of Lola show that even so far as there is real "intelligence" in the animal, the supernormal relationship enters very quickly on the scene. In other words, the subject very quickly learns to express itself by means of a true "xenoglossy," i.e. by means of a language that may be clear to other people although it probably is not understood by the animal or medium making use of it.

Besides, we find in Lola's case a high degree of glossolalia. The authoress observes, e.g. on page 39: "Lola often uses words completely incomprehensible; at one time she declared that they belonged to a special canine language. My investigations on this subject remained, however, without result. It is possible that these words arise from the imagination of the animal...."[30] Something similar was also produced by Rolf and the Elberfeld horses.

Of course, even after the development of this "xenoglossy," it is difficult either to admit or to refuse to admit some remainder of self-conscious co-operation by the animal in its "answers." For my part, I believe that simple replies may continue to be formed in the normal self-conscious way. It is certain, in my opinion, that this view is one of the only two alternatives possible when we get replies to questions the contents of which are entirely unknown to everybody else present. The other alternative is that of clairvoyance in those present followed by projection by them to the animal of the idea obtained clairvoyantly; or else of a "telepathic" projection of the sense-impression from the animal to the bystanders, with return of the reply from the latter to the former. I do not dare to complicate this further; the more so as in all the cases which I know of in which replies were obtained to such questions, very simple things only were dealt with: figures, or modest problems; or else problems which are abstruse "to us," such as fourth and fifth roots, but which as the animal was one of the horses at Elberfeld may be explained by the general mathematical faculty without drawing upon the mediumistic hypothesis.

But that there is on the whole much of the subliminal at work in all the cases noted is, I believe, difficult to deny.

We must remember that superior "force" by which Miss Kindermann felt herself, as it were, compelled (page 36). And in another place (page 40), the authoress declares: "However strange it may seem, I have repeatedly remarked that Lola always finds abstract calculation and spelling easy; whilst on the other hand it always seems difficult to make her move single parts of her body, or to carry out practical orders." (I myself was able to make similar observations at Elberfeld and at Mannheim; it seemed to me, however, that the horses were more docile to "practical orders.").

On page 42 I find: "During the explanation of the digits and of the tens, the dog did not look at me, but bit with apparently very great interest a leg of the stool." It must be noted, as I have already pointed out, that the digits and the tens were both alike learned quickly and well. The authoress explains this action of Lola's as a "mark of embarrassment." But to me that leg of the stool is exactly on a par with the salad leaf mentioned by Professor Ferrari: i.e. the dog did not pay the slightest attention to the lesson; it replied without the help of intelligent attention on its part; it replied in the subliminal way, like the unconscious instrument of a psychic automatism, and by the use of an intelligence which was not its own.

Similar impressions are left by other points in the story of Lola. I read on page 64: "If, for instance, I write one under the other three or four numbers of two figures each, very quickly, and without adding them myself, and then hold up the sheet in front of the dog, I see that her eyes only glance at the sheet for 1-2 seconds; after which the dog bends its head to add but looks away, and then taps the reply." This behaviour is the same as that of Krall's pony Hanschen, when Dr. Assagioli and I made experiments with it.

The same can be said of various other performances of an intuitive kind, on the part of Lola, to which the authoress refers: e.g. knowledge in four seconds of a given number of points (up to 35), marked without any regularity whatever on a piece of paper. (Similar experiments were made at Elberfeld and Mannheim.) Other performances of an intuitive kind concern various measures of time, temperature, musical intervals, etc., and they reach their highest point in the *premonitions* as to the course of the weather and the birth of the puppies. Professor Ziegler finds the explanation of this last performance in the prenatal movements of the fœtus within the maternal body. This seems to me doubtful; besides, it must be remembered that this prevision of Lola's was a double one, as it concerned both the number and the sex of the puppies (autoscopia?). The fact that the sex of the puppies was foretold *almost* correctly does not eliminate all doubt. And the authoress gives sufficient details on the experiment to make us regard it as genuine, until we have proof to the contrary.

Many other manifestations of Lola's betray very clearly a subliminal relationship between herself and her mistress (or perhaps between herself and other persons), and so I do not see that there is any reason for us to doubt that Miss Kindermann was really surprised at the replies which she obtained.

I could cite at length: I am content, however, to remind the reader of the many replies of the dog which reveal quite clearly the feeling of the authoress towards the dog itself, as e.g., "I know you, alas, so little"; or again, "Show constancy in your love for me," etc.; then, again, the words never pronounced before in presence of the dog (this makes me think of the famous "Urseele" of Rolf); the things said by Lola, but not known by her mistress, and then found true....

Finally I must allude to the "discovery" made by Lola that the odours of the human body reveal the state of the human mind — displeasure, jealousy, lie (sic); on which the authoress observes (very justly, in a certain sense), that these experiments make one think of the well-known theories of the late Prof. Jaegar of Stockholm.... I am in agreement with her on that point, because I, too, have read the "Entdeckung der Seele" by that author; as I suppose she, too, had. I am inclined to think that in her case (as she was experimenting with a dog) it was only natural for her to think of these psycho-olfactory theories — perhaps without knowing it — even before the experiments. Therefore, the experiments themselves would always be perfectly "genuine," but of course this genuineness is of a different sort to what she thought it.

To conclude, the supporters of the new zoopsychology must not complain if the views which I have set out above help in course of time to oust their "point of view." It seems to me that even while robbing the "thinking" animals of some of the intelligence attributed to them, and while regarding what remains as qualitatively different from human intelligence (e.g. through the much greater interference of subconscious factors), we are still free to find the animals to be perhaps even more interesting than before.

I am quite conscious of the fact that the "cases" are still few for theories to be built upon; and some may think that I might have done better by reporting them simply without attempting any explanation whatever. However, I believe, that if as the result of my work the recognition of the internal weakness of certain hypotheses — especially in the psychological field — is generally recognized, it will not be so harmful to have put forward some suggestions for dealing with facts which have already been, or will be, established.

I have accordingly tried to do so, but I shall always be ready to modify my views if new facts should persuade me that this is necessary.

Postscript. — Professor G. C. Ferrari has published an article on Lola in *Rivista de Psicologia*, 1920, 1. His explanation corresponds in many points with my own.

BIBLIOGRAPHY

1. E. Bozzano. "Animals and Psychic Perception." *The Annals of Psychical Science*, II, 2. London, 1905.

2. E. Morselli. *Psychology and Spiritism* (in Italian). Fr. Bocca, Turin, 1908.

3. C. de Vesme. "The Thinking Horses of Elberfeld" (in French). *Annales des Sciences Psychiques*, XXII, 12. Paris, 1910.

4. G. C. Ferrari and F. Pulle. "The First Month of a Horse's Education" (in Italian). *Rivista di Psicologia*, March-April, 1913.

5. E. C. Sanford. "Psychic Research in the Animal Field." *American Journal of Psychology*, XXV, 1914.

6. P. Sarasin. "Animal and Human Quick Reckoners" (in German), from *Proceedings of the Natural History Society at Basle*. Basle, 1915.

7. H. E. Ziegler. *The Soul of the Animal* (in German). W. Junk, Berlin, 1916.

8. W. Neumann. "Pseudo-animal-psychology" (in German), from *Naturwiss: Wochenschrift*. Jena, 1916.

9. W. Mackenzie. *At the Founts of Life* (in Italian). A. F. Formiggini, Rome, 1916. (Out of print).

10. W. Maeterlinck. *The Unknown Guest*. Methuen, London.

11. P. Moekel. *My Dog Rolf* (in German). R. Lutz, Stuttgart, 1919.

12. W. Mackenzie. "Rolf of Mannheim." Translated by Miss E. Lathan, with notes by Professor J. H. Hyslop. *Proceedings of the American Society for Psychical Research*. New York, August, 1919.

13. H. E. Ziegler. "The Memory of the Dog" (in German), from *Zoologischer Anzeiger*. Leipzig, November, 1919.

14. H. Kindermann. *Lola*. Jordan, Stuttgart, 1919.

15. G. C. Ferrari. "What Talking Dogs Think" (in Italian). *Riv. di Psicologia*, Bologna, XVI, 1. 1920.

Footnotes

[1] Published by Friedrich Engelmann, Leipzig.

[2] Published by the committee through the agency of Professor Ziegler.

[3] Published by Emil Eisell, in Bonn.

[4] Frau Dr. Moekel told me that she again asked the dog on the following day what the article shown him had been and he answered: "hd sdld bei arm grosfadr grab lib maibliml" (Hat gestehlt bei des armen Grossvaters Grab das liebe Maiblümchen) (Had stolen from dear grandfather's grave the dear little lilies-of-the-valley!). The object shown him had been a lily-of-the-valley, and a few days before, Frau Moekel's mother had told the children that she had taken all the lilies-of-the-valley to their grandfather's grave. Rolf, therefore, seemed to have conceived the idea that the flowers shown him had been pilfered. — Ziegler.

[5] The hatred of dogs for cats is hereditary; it is an instinct common to all dogs, and, seeing that instinctive sensations do not owe their origin to any deliberate act of reasoning, it is generally difficult to account for them. It is therefore worth drawing attention to the fact that Rolf did, nevertheless, make an attempt at giving a reasonable reply. — Ziegler.

[6] Taken from the "Communications of The Society for Animal Psychology," 1916. pp. 6-9.

⁷ These dogs were born on 26 and 27 January, 1914. Compare the letter of Rolf in the "Communications of the Society for Animal Psychology," 1914, p. 28; and "The Soul of the Animal," p. 111.

⁸ Ilse was barely two months old when she came into the possession of her master, on 20 April, 1914.

⁹ The dog had become familiar with square roots in the course of earlier attempts.

¹⁰ Frau Dr. Moekel taught another young dog, called Lux, as well as Roland, the former being taken over by a gentleman in Mannheim. In a protocol dated 14 June, 1914, I stated that Lux was able to do a certain amount of arithmetic at the age of four and a half months.

¹¹ Professor Karl Kindermann, of Hohenheim.

¹² Gegs = keks; Germans call biscuits "keks."

¹³ Here observe that Rolf has the impudence to complain of the Moekels for not feeding him on sweet biscuits!

¹⁴ So as to avoid confusing her I always write the *sound* only.

¹⁵ *h* is the term used in Germany for the note we call *b*.

¹⁶ Maulburg, near Schopfheim, in Baden, where Lola had visited relations of mine.

¹⁷ Mittel = unbestimmt (uncertain; from Mitte = middle.)

[18] Fractions will be touched on in a later chapter on "Advanced Arithmetic."

[19] Chapter XVIII, "Spontaneous Answers."

[20] The poet, Hans Müller, has touched most eloquently on the power to think latent in animals in his book, "Die Kunst sich zu freuen."

[21] At a meeting held by the Rolf Society at Stuttgart, Professor Ziegler accounted for this accurate knowledge by declaring that—prior to birth—the puppies lie in a row within their mother's womb, and that if one moves, the others proceed to move also, but only one after the other.

[22] Sie = you is the more formal mode of address, as opposed to the familiar "du" = "thou."

[23] Lola often uses quite incomprehensible words and once declared that they belonged to "a particular dog-language"—my further inquiries have been quite fruitless, and these words were probably her own inventions!

[24] "Who are you?"

[25] See the Song of Solomon.

[26] I would here refer the reader to the references I made to the work issued by Pfungst; they may be found in "The Animal Soul" (Reports of new observations made with respect to horses and dogs), 2nd ed. (W. Jung) 1916, p. 38.

[27] Karl Krall, "Denkende Tiere, Beiträge zur Tierseelenkunde, auf Grund eigener Versuche," Leipzig, Engelmann, 1912.

[28] Rolf could only rap with one paw owing to the other fore-paw having been injured; he generally leaves out the vowels, these being already contained within the consonants. This habit gives rise to a somewhat curious form of writing.

[29] Note.—The numbers in the text refer to the Bibliography at the end.

[30] N.B.—It may also be that the "quite incomprehensible words" have not any meaning at all, or at least, not any relation with the mechanism of the glossolalia, but are simply the product of taps made by the animal just for the sake of doing something.

www.ingramcontent.com/pod-product-compliance
Lightning Source LLC
Chambersburg PA
CBHW031416210526
45464CB00005B/1907